THE AI READY CLASSROOM:

Preparing Educators for the Future of Learning

Matthew J. Welker, Ed.D.

i

Copyright Notice

Copyright © 2026 Matthew J. Welker. All rights reserved.

No part of this publication may be reproduced, stored in a retrieval system, or transmitted in any form or by any means, including electronic, mechanical, photocopying, recording, or otherwise, without the prior written permission of the copyright owner, except as permitted by U.S. copyright law (including fair use) and for brief quotations in reviews and critical articles.

For permissions requests, contact: welker@meded-consulting.com

Acknowledgments

I am grateful to the people who made this book possible.

First and foremost, I want to thank my wife, Charlene, for her unwavering love, patience, and encouragement throughout this journey. Her support gave me the clarity and motivation to see this project through every challenge and late-night revision.

I am also deeply grateful to the students and teachers of Miami-Dade County Public Schools. Your curiosity, innovation, and commitment to learning have been a constant source of inspiration for me throughout my career. The stories, insights, and real-world experiences you shared with me helped shape this book's vision of how technology, and especially artificial intelligence, can empower education.

Finally, to all educators and lifelong learners exploring new ways to teach and grow in the age of AI: this work is for you.

AI can be a powerful tutor or a subtle crutch; its value in education is determined not by what it knows, but by how wisely educators ask students to use it.

Table of Contents:

Introduction:
Teaching and Learning in the Age of Artificial Intelligence (AI)

Artificial intelligence (AI) has moved into education with unusual speed. Tools that once sat at the margins, such as automated scoring, adaptive tutoring, and predictive dashboards, now shape how schools plan, teach, and assess. With the rise of large language models and other generative systems, AI has expanded beyond calculation and pattern recognition into work that resembles parts of human reasoning: explaining ideas, drafting instructional materials, imitating student writing, and responding to learner questions in real time.

For teachers, this shift brings both opportunity and uncertainty. AI can generate lesson ideas, reduce administrative load, support differentiation, and provide forms of just-in-time assistance that time rarely permits. For students, it can serve as an always-available tutor, clarifying misconceptions, modeling reasoning, and providing language support for multilingual learners. In districts facing staffing shortages, widening achievement gaps, and rising expectations, AI can feel less like an innovation and more like a necessity.

Yet the promise of AI masks a deeper issue that runs through this book: What happens to the human work of education when machines begin to interpret, recommend, and generate knowledge in the classroom? AI can expand capacity, but it also shifts the boundaries of professional judgment. It raises urgent questions about what should be automated and what must remain human; what counts as learning and who decides; how equity is advanced or undermined; and what kind of person education seeks to cultivate. The choices educators and policymakers make now will determine whether AI widens opportunity and inclusion or narrows education by displacing human wisdom and compassion.

This book is written for teachers, school leaders, college administrators, and education policymakers. Its central claim is simple: teaching must remain a human practice, grounded in ethical intention, meaningful relationships, and careful attention to context. AI can process vast amounts of data, but it cannot hear the emotional undertone in a child's voice, recognize the meaning of cultural nuance, or weigh the ethical implications of a decision. It cannot offer the encouragement that restores a struggling learner's confidence or the wisdom that situates knowledge within a larger moral horizon. These capacities, trust, empathy, and discernment, are not extras. They are the heart of learning.

The chapters that follow show that teachers' roles will not remain static. As AI enters daily practice, educators are pushed toward expanded work: designing learning experiences and environments rather than simply delivering content; coaching students in inquiry, metacognition, and resilience; and serving as ethical stewards who set boundaries and supervise powerful tools with care. Students, too, need new literacies so they can question machine outputs, recognize limitations, and collaborate with AI without surrendering agency or intellectual integrity.

This book also confronts the structural realities surrounding AI adoption. AI systems are not neutral; they are built within social, political, and economic conditions that shape how they work and whom they serve. Without clear governance, AI can reproduce historical inequities, concentrate power in vendors, or reduce complex learners to data points stripped of context. That is why implementation must be guided by policy, transparency, privacy protections, and explicit equity commitments, not as bureaucratic hurdles, but as ethical guardrails that protect student dignity, strengthen trust among educators and families, and preserve public confidence in schools.

This is neither a case for resisting AI nor a manifesto for embracing it uncritically. It is a guide for leading education through one of its most consequential technological shifts since the rise of mass schooling. The lens throughout is teaching's deepest purpose: the formation of flourishing human beings. When guided by clarity, humility, and a steadfast commitment to humanity, AI can amplify teacher expertise and broaden student opportunity. The goal is not to substitute for educators and learners, but to strengthen their work so technology becomes a tool for empowerment, creativity, and meaningful engagement rather than a replacement for human connection and wisdom.

By the end of this book, readers will have a conceptual framework and a practical toolkit for engaging AI responsibly, protecting teacher autonomy, strengthening student agency, confronting algorithmic bias, and supporting learning that is equitable, ethical, and relational. Through case studies, reflective questions, and actionable strategies, readers will be prepared to navigate the complexity of AI integration without losing sight of what matters most: human beings at the center as designers, guides, and stewards of learning.

In the age of AI, the central challenge is to ensure that technology magnifies human wisdom rather than displaces it. The future of education will not be determined by AI's power alone, but by the courage, judgment, and moral imagination of those who teach with it. Educators, policymakers, and learners share responsibility for shaping AI use in ways that foster justice, empathy, and deep learning, and for ensuring that the next generation inherits not only advanced tools, but a more thoughtful and caring society.

Part I:
The AI Revolution

Chapter 1:
The Impact of Technology on Education

Education has always been shaped by the tools societies use, but in schools, tools never arrive "just" as tools. They change what is easiest to teach, what students are expected to know, what gets assessed, and how teachers spend their time. For educators today, understanding earlier technology shifts is not an academic exercise; it is a practical way to make better decisions about AI adoption. The central lesson of history is this: when technology changes, instruction, assessment, and equity pressures change with it (Postman, 1985).

This chapter offers that perspective by tracing a long arc, from early cognitive tools like the abacus and writing systems, to the printing press, to industrial-era classroom devices, to computers, the internet, and today's AI systems. The goal is not nostalgia; it is pattern recognition. Across eras, the same questions reappear in new forms: What becomes easier to teach and measure? What kinds of thinking are strengthened or weakened? Who gains access and who is left behind? And how does technology shift the locus of authority in classrooms and systems? Keeping these questions visible from the start prepares the reader to engage the rest of the book with clarity about what is at stake when AI begins to participate in the work of teaching and learning.

Understanding the history of educational technology is essential to assessing current developments in artificial intelligence. Current debates about automation, algorithmic bias, and digital learning echo earlier anxieties about new educational technologies such as the printing press, slide rule, and handheld calculator. Each invention has raised both hope and unease, transforming not only instructional methods but also the policies, economies, and moral assumptions that undergird schooling (Saettler, 2004). What changes in the AI era is the proximity of the tool to judgment: AI does not merely deliver information; it can generate, interpret, and recommend, which means it can shape feedback, assessment, and opportunity if educators are not deliberate. Reading today's debates through history helps educators separate familiar patterns of disruption from genuinely new risks, especially around equity, transparency, and what must remain unmistakably human.

This chapter also traces the evolution of educational technology from ancient tools to artificial intelligence to show a recurring pattern: tools expand access and efficiency, but they also reshape what schools value and how students are judged. Placing AI in this historical lineage helps educators anticipate predictable implementation challenges around assessment, equity, and professional judgment (Eisenstein, 1979; OECD, 2025). The core argument is that AI is not simply a faster resource; it is a technology that participates in cognition (generating, interpreting, and recommending), which raises higher-stakes questions about what must remain human. With that foundation in place, the chapters that follow move from history to present practice: Part I establishes why AI represents a new phase of educational technology, and the later parts turn to teacher preparation, instruction, assessment, and policy so readers can apply these historical lessons to real decisions.

The Ancient Foundations of Learning Technologies

The history of educational technology begins not with computers or books but with the abacus, one of humanity's earliest cognitive tools. Archaeological evidence traces primitive counting boards as far back as 2300 BCE in Mesopotamia, where merchants and scribes used rows of stones or beads to record transactions (Ifrah, 2000). Later refinements in China, Greece, and Rome gave rise to the *suanpan* and *abacus*, which formalized numerical reasoning and introduced students to the concept of abstract representation (Robson, 2008). The abacus illustrates an enduring theme in education: when a tool makes thinking visible, it can accelerate learning, but it can also shift what we practice and what we expect students to do mentally. That same tension reappears later with calculators, computers, and now AI.

The next leap in educational technology was the invention of writing systems and the materials that supported them: clay tablets, styluses, papyrus scrolls, and later parchment. Writing extended human memory and stabilized knowledge beyond the moment of speech, transforming the teacher's role from oral transmitter to interpreter of text (Robson, 2008). In early schooling contexts, this shift also changed what counted as evidence of learning: durable inscriptions could be reviewed, compared, and corrected, making student work more visible and more permanently evaluable than oral performance. It also introduced a new instructional practice: repetition through copying. Students learned literacy by inscribing signs repeatedly, developing fluency through guided imitation and correction, an early analog to the iterative learning reinforced today by digital practice systems. In both cases, the tool supports scale and consistency, while subtly redefining literacy as a skill demonstrated through standardized artifacts.

Writing technology expanded the scale of instruction by enabling knowledge to be recorded, preserved, and transported across both time and distance. This development allowed scholars, teachers, and learners not only to access a growing body of accumulated wisdom but also to build upon the insights and discoveries of others, fostering intellectual progress across generations. In Plato's *Phaedrus*, Socrates famously criticized writing for weakening memory and undermining authentic dialogue, expressing concerns that written words might replace the deeper internalization and understanding achieved through face-to-face discourse. This ancient debate about the risks and rewards of technological mediation (a tension between technological augmentation and human authenticity) echoes in later and contemporary critiques of calculators, computers, and AI chatbots as well (Postman, 1985). The struggle to balance the benefits of innovation with the preservation of core human capacities is a recurring theme throughout the history of educational technology.

The Printing Revolution and the Democratization of Knowledge

The invention of the printing press by Johannes Gutenberg around 1440 marked one of the most important developments in the history of education. By mechanizing the reproduction of text, the press drastically reduced the cost and time required to produce books, enabling a wider dissemination of knowledge than ever before. Eisenstein (1979) argued that this innovation "reoriented the conditions of intellectual life," making possible the mass literacy movements of the Renaissance and Reformation. The accessibility of printed materials encouraged not only the spread of new ideas but also the standardization of curricula, paving the way for the development of formal education systems and expanding opportunities for learning across social classes.

Before printing, education was limited largely to elites with access to hand-copied manuscripts. After printing, textbooks and primers became affordable for the emerging middle classes. The printing press democratized knowledge, standardized spelling and grammar, and enabled shared curricula, laying the foundation for national education systems (Eisenstein, 1979). Its effects extended beyond schooling. Governments and religious institutions quickly recognized the political power of printed knowledge. State-sponsored schools used standardized text materials to shape civic identity, while reformers used pamphlets to challenge authority. Education policy thus became linked to information policy, a dynamic that continues in debates over digital content regulation and AI-generated materials.

By the 16th and 17th centuries, the printed book had become the central artifact of education. Classrooms were redesigned around reading and recitation, and teachers evolved from oral performers into interpreters of print. The ratio of students to materials changed. Instead of one teacher reading to many listeners, many readers could now study individually or in small groups. Printing also introduced the notion of standardized assessment. With identical copies of texts, teachers could assign the same passages or problems, paving the way for formal testing and credentialing (Saettler, 2004). This shift from apprenticeship to institutional schooling represented a significant technological and cultural transformation, one that remains with us today.

In policy terms, mass print education aligned with the rise of nation-states and industrial economies. Literacy became both a civic duty and a labor requirement. As the Organisation for Economic Co-operation and Development (OECD) (2025) notes, technological change is often accompanied by corresponding shifts in what education systems value, govern, and teach. In this case, the press catalyzed bureaucratic systems for managing curriculum, certification, and teacher training.

The printing revolution parallels today's digital transformation in important ways. Both eras expanded access to information dramatically, destabilized existing hierarchies, and provoked moral panics about authenticity and authorship. Just as critics once feared that cheap print would erode intellectual discipline, modern educators worry that artificial intelligence will weaken students' reasoning and originality (Selwyn, 2019). History suggests, however, that technology does not inevitably diminish learning; rather, it demands new literacies. The transition from manuscript to print required stronger reading comprehension and critical interpretation, skills that then became newly central. Similarly, the age of AI requires new competencies in algorithmic literacy, data ethics, and human–machine collaboration (UNESCO, 2024).

The printing press reminds us that educational technologies are disruptive but also generative. Likewise, the rise of AI challenges traditional concepts of literacy, assessment, and educational equity, demanding that we reconsider not only how information is delivered but also how learners engage with and interpret it. By drawing lessons from earlier transitions, such as the shift from oral to written culture and from manuscript to print, today's stakeholders can shape AI's integration in ways that foster inclusion, critical thinking, and lifelong learning rather than amplify existing divides.

Although separated by five centuries, the logic of the printing press underpins today's digital infrastructure. Both depend on reproducibility, distribution, and scalability that make knowledge widely available. Yet each new medium also raises questions about control and inequality. In Gutenberg's time, literacy expanded but also stratified populations between readers and non-readers. In the AI era, digital fluency and access to data likewise shape who benefits most from innovation (OECD, 2025).

Modern educators, like those before them, seek to use learning tools in ways that expand access and uphold educational purpose. The printing press once shifted education toward the written word, while AI is now evolving teaching into a more interactive and flexible experience where teachers help students interpret and understand, not just deliver facts (Holmes et al., 2019). Despite these changes, one fundamental question remains: How can technology enhance educational goals without compromising them? Most earlier classroom technologies primarily helped educators store, display, copy, or distribute information. AI is different because it can generate (create text and tasks), interpret (summarize and explain), and recommend (predict needs and suggest actions). That shift matters because it moves technology closer to work that has traditionally depended on teacher judgment, especially feedback, assessment, and instructional decision-making.

The Age of Mechanical and Analog Devices (1600s–1800s)

Though deceptively simple, the pencil is one of the most important tools in educational history. Developed in the late 16th century after the discovery of graphite deposits in Borrowdale, England, the pencil became a portable, affordable, and erasable instrument for recording and revising thoughts (Petroski, 1990). Its educational significance was substantial. Unlike ink pens, which required precision and permanence, pencils encouraged experimentation and iterative thinking. Students could work their way toward understanding through drafting, revising, and visualizing ideas. This cognitive process is closely aligned with what later became known as design thinking and formative assessment.

By the 19th century, mass production of pencils by companies such as Faber and Dixon made writing accessible to children worldwide. In educational terms, this democratization of expression paralleled the printing press's democratization of access. Like later computing tools, the pencil lowered the cost of intellectual participation. It symbolized the shift from elite scholarship to universal education (Saettler, 2004). In policy history, the pencil's spread coincided with compulsory schooling laws in Europe and North America, illustrating how simple technologies often underpin large-scale social reform.

Invented by William Oughtred in the early 1600s, the slide rule represented an early mechanical method for performing rapid mathematical calculations (Ifrah, 2000). This ingenious device made use of logarithmic scales, allowing users to multiply, divide, and solve complex equations by simply sliding parts of the rule against each other. For centuries, engineers, navigators, and students relied on the slide rule to compute products, quotients, and trigonometric functions that previously required extensive manual calculation. Historical documentation reinforces its central place in scientific and engineering practice before the rise of electronic computation, underscoring its role as a foundational tool in both educational and professional settings (Oughtred Society, 2013).

Educationally, the slide rule represented a fusion of mathematical reasoning and mechanical ingenuity. Mastery of it required not rote memorization but conceptual understanding of ratios and exponents. Its instructional use peaked during the 20th century, particularly in engineering education, before being supplanted by electronic calculators in the 1970s (Saettler, 2004). The slide rule familiarized students with early computational thinking by showing that problems could be approached algorithmically through the systematic use of tools. It foreshadowed later developments in computer programming and algorithmic logic (Denning & Martell, 2015).

Industrial-Era Technologies (Late 1800s–1950s)

As education expanded in the 19th and early 20th centuries, teachers faced a new challenge: how to deliver consistent instruction to large groups of students efficiently. The mimeograph, patented by Thomas Edison in 1876, revolutionized classroom logistics by enabling teachers to duplicate worksheets, exams, and study materials quickly (Saettler, 2004). Before digital printers, the mimeograph became a central tool of classroom communication: an analog forerunner of today's word processors and learning management systems. It allowed the standardization of materials, which in turn supported the growth of national curricula and assessment frameworks.

A generation later, the overhead projector (introduced widely after World War II) transformed visual instruction. Transparent acetate sheets could be written on, erased, and reused, making lessons interactive in a pre-digital way. According to Saettler (2004), these tools "shifted classroom attention from the printed page to the visual field," encouraging multimodal learning. While seemingly mundane, these technologies democratized access to information and transformed pedagogy from teacher-centered lecturing to visual and participatory learning, a shift echoed today in AI-powered interactive platforms.

Broadcast and Digital Eras (1950s–1980s)

The mid-20th century introduced an era of audiovisual learning, in which motion pictures, radio, and television were harnessed to bring the world into the classroom. During the 1950s and 1960s, educational television initiatives proliferated in the United States and abroad, often supported by public funding and UNESCO partnerships (Cuban, 1986). Television changed teaching by using images, sound, and narrative to make complex ideas more accessible. Students could watch scientific demonstrations, historical reenactments, or language lessons in real-time, broadening cultural horizons.

However, as Postman (1985) observed, the medium also introduced concerns about passivity and entertainment-driven learning. The television classroom risked turning education into a spectacle rather than inquiry. This critique anticipated later concerns about digital media consumption, screen time, and attention fragmentation (Selwyn, 2019). However, the audiovisual era marked a major step toward mass media literacy. Teachers learned to integrate filmstrips and TV broadcasts into lessons, and instructional designers began developing early models of media-based pedagogy, a precursor to today's multimedia and blended learning designs.

When the first handheld calculators became affordable in the 1970s, they triggered intense debate among educators. Detractors claimed that calculators would weaken mental computation, impede the development of foundational problem-solving skills, and encourage overreliance on the device. Proponents argued that they freed learners to focus on problem-solving and higher-order reasoning.

This debate resembles current discussions about AI tools such as ChatGPT and adaptive tutors (OECD, 2025). In both cases, educators had to reimagine assessment: Was it fair to test manual skills when technology could perform them instantly?

Research from the late 1970s and 1980s found that, when used appropriately, calculators improved conceptual understanding and motivation (Ellington, 2003). They also helped redefine mathematical literacy as the ability not merely to compute by hand, but to understand and apply quantitative reasoning. The calculator era underscored a recurring pattern: when a cognitive tool automates one task, education shifts its focus to higher-order work. Just as calculators displaced arithmetic drills, AI now challenges educators to emphasize creativity, judgment, and ethical reasoning (Holmes et al., 2019).

The Computer Age and Digital Literacy (1980s–2000s)

The introduction of the personal computer in the 1980s marked a major shift in educational history. For the first time, learners could interact with digital environments individually rather than collectively. Early educational software such as Logo (Papert, 1980) and Reader Rabbit, introduced in 1984, exemplified the potential for constructivist learning through interaction. Papert's theory of constructionism argues that students learn best by building tangible artifacts like programs, stories, and simulations with computers, shifting education from passive reception to active, hands-on creation. This idea helped reshape educational practice by emphasizing learning through making rather than passive content absorption.

Policy initiatives soon followed. Programs such as "Computers in the Classroom," introduced by the U.S. Department of Education in 1983, and the "BBC Micro" project in the United Kingdom sought to integrate computing into every school (Cuban, 1986). By the late 1990s, computer literacy had become a core competency, viewed as essential as reading and writing. The PC revolution also intensified concerns about digital equity. Access to computers often correlates with socioeconomic status, creating a "digital divide" that policymakers continue to address. This divide, once defined by hardware access, now includes disparities in data, software, and algorithmic literacy (UNESCO, 2024).

The educational importance of coding emerged alongside the rise of the personal computer. Early programming languages such as BASIC, introduced in 1964, Pascal, introduced in 1970, and Python, introduced in 1991, made it possible for students to write instructions that machines could execute. This shift marked a transition from consuming information to creating computational systems (Denning & Martell, 2015). Coding introduced learners to fundamental concepts of logic, abstraction, and problem decomposition; skills now recognized as essential for all disciplines. Jeannette Wing's influential 2006 essay suggested that programming goes beyond technical expertise; she claimed it is a universal way of thinking that can be applied to any discipline (Wing, 2006).

From an educational policy perspective, nations have since integrated coding into K–12 curricula, recognizing it as a core literacy of the 21st century (OECD, 2025). However, this movement also highlights equity concerns: students in under-resourced schools often have less access to quality instruction in coding and robotics, reinforcing digital hierarchies (UNESCO, 2024). The development of programming languages, from assembly code to Python, parallels broader shifts in human communication tools. Just as writing enabled humans to externalize thought, programming languages externalize processes of reasoning, enabling collaboration between humans and machines.

Although AI research dates to the 1950s, its educational applications have accelerated primarily over the last two decades. Early systems, such as the PLATO project (Programmed Logic for Automated Teaching Operations) at the University of Illinois in the 1960s, used mainframe computers to deliver customized instruction (Suppes, 1966). These early intelligent tutoring systems prefigured the adaptive platforms of today, such as Khan Academy's mastery learning models and AI-driven assessment systems. Modern AI tools analyze vast amounts of learner data to personalize instruction, identify gaps, and even generate practice problems or feedback (Holmes et al., 2019). These tools promise to make education more adaptive and inclusive, but they also raise concerns about privacy, bias, and the erosion of teacher autonomy (Mobilio & Guglielmini, 2026). AI continues the long trajectory begun with the abacus and printing press: each new "thinking tool" redefines the human role in learning.

By the mid-1990s, the internet became a groundbreaking educational tool, rivaling the impact of the printing press. Students gained unprecedented access to vast resources and could collaborate with others worldwide for the first time. The World Wide Web, developed by Tim Berners-Lee in 1989, evolved from an information-sharing project into a universal educational platform (CERN, 1993; Berners-Lee, 1999). The internet made information more accessible than earlier broadcast media ever had. Online encyclopedias, digital libraries, and later Wikipedia (founded in 2001) offered open access to knowledge once restricted to academic elites (Reagle, 2010). Distance education, which had existed since the era of correspondence courses, expanded exponentially through early online platforms such as Blackboard, launched in 1997, and Moodle, launched in 2002.

Within the educational environment, the internet ushered in the era of constructivist learning networks. Students shifted from relying primarily on textbooks and instructors to locating, evaluating, and synthesizing information for themselves. The teacher's role shifted toward that of a curator and guide: a transformation that is echoed today in AI-supported classrooms (Selwyn, 2019; OECD, 2025).

However, the internet also magnified inequalities. Bandwidth, device access, and digital literacy created new divides between connected and unconnected populations (UNESCO, 2024). These disparities persist, illustrating that technological innovation alone cannot produce educational equity without thoughtful policy.

The Interactive Era (2000s–2010s)

The introduction of interactive whiteboards, also referred to as smartboards, in the late 1990s and early 2000s represented the convergence of tactile, visual, and digital learning. Teachers could display multimedia lessons, annotate in real-time, and invite students to manipulate content directly on the board. According to Higgins et al. (2007), smartboards enhanced student engagement by transforming the classroom from a static environment into an interactive learning space. Students could explore simulations, collaborate on digital maps, or analyze data visually, which were precursors to today's interactive AI platforms.

Smartboards also marked a significant instructional shift from content delivery to content co-creation. The traditional teacher-led lecture evolved into a dialogic, multi-sensory experience. Importantly, these technologies supported Universal Design for Learning (UDL) principles, helping accommodate students with diverse learning styles and accessibility needs (Rose & Meyer, 2006). However, their success depended heavily on teacher training and institutional support. Without proper professional development, schools often use interactive whiteboards as "expensive chalkboards," demonstrating that technological adoption requires instructional transformation to be effective (Cuban, 2013). The proliferation of smartphones and tablets in the 2010s marked the dawn of ubiquitous learning (u-learning), which made education accessible anytime, anywhere. Mobile devices blurred the line between formal and informal learning environments, enabling students to watch tutorials, participate in online discussions, or conduct research from their pockets (Traxler, 2009).

Apps such as Duolingo, Khan Academy, and Google Classroom revolutionized learning accessibility, offering personalized, gamified experiences for millions of users worldwide. These tools embody the constructivist and self-directed learning ideals envisioned by Dewey and Papert, empowering students to pursue knowledge autonomously. The mobile era introduced new ethical and cognitive concerns. Constant connectivity risked fragmenting attention spans, while algorithm-driven content raised questions about privacy and data exploitation (Selwyn, 2019). Schools were forced to balance innovation with digital well-being and screen-time management policies. The mobile revolution thus reinforced a familiar pattern: every technology that expands access to learning also challenges educators to redefine balance, focus, and purpose.

Artificial Intelligence and the Cognitive Revolution (2010s–2020s)

The 2010s witnessed the rapid expansion of Massive Open Online Courses (MOOCs) and open educational resources (OERs). Platforms such as Coursera, edX, and FutureLearn promised to make world-class instruction freely available to anyone with an internet connection (Pappano, 2012). MOOCs and OERs epitomized the open knowledge movement, a digital echo of Gutenberg's democratization of print. They allowed learners from developing regions to access elite university courses, fostering global academic collaboration. However, critics noted that completion rates remained low and participation often skewed toward already-educated individuals (Hollands & Tirthali, 2014). The promise of universal education met the realities of motivation, infrastructure, and cultural context. Nevertheless, open education transformed institutional practices. Universities began adopting blended learning models, combining online and face-to-face instruction, and policymakers recognized OERs as cost-effective tools for equity (UNESCO, 2024).

Artificial intelligence, once a theoretical concept, now permeates nearly every aspect of education, from adaptive learning platforms and automated grading to personalized tutoring and predictive analytics. Modern AI systems can diagnose student misconceptions, suggest tailored exercises, and even generate essays or problem sets (Holmes et al., 2019). AI represents not just another tool but a new cognitive partner. Learning systems can adapt to each student's progress in real-time, offering tailored support much like Vygotsky's zone of proximal development, serving as a kind of algorithmic scaffolding (Holmes et al., 2019; Vygotsky, 1978).

AI applications such as chatbots, translation tools, and learning analytics dashboards promise unprecedented efficiency and inclusion. For instance, speech-to-text and translation AI can enhance accessibility for multilingual learners and students with disabilities (UNESCO, 2024). However, AI also introduces significant ethical and epistemological challenges. Data privacy, algorithmic bias, and overreliance on automated systems can distort educational fairness and weaken human agency (Selwyn, 2019). Large language models may generate plausible but inaccurate information, requiring new forms of critical AI literacy and stronger attention to ethical use, human oversight, and professional judgment (UNESCO, 2024). As OECD (2025) notes, educators must balance innovation with integrity, ensuring that AI complements rather than replaces human judgment. Teacher preparation programs must now incorporate not only digital pedagogy but also data ethics and algorithmic transparency.

The ultimate trajectory of technology in education points toward augmentation, not automation. Future classrooms will likely integrate AI seamlessly with human instruction using predictive analytics to guide teachers, generative models to provide scaffolds, and virtual agents to support emotional well-being. In this future, the teacher becomes a learning architect, orchestrating human and artificial intelligences toward shared goals (Holmes et al., 2019). Students, meanwhile, develop metacognitive and ethical fluency, learning not only how to use AI but when and why to use it.

Policymakers play a vital role in this transition. UNESCO published a policy framework emphasizing the need for ethical, human-guided use of AI. It also argued that AI policies developed for schools must safeguard privacy, ensure transparency, and promote equitable access. Without such frameworks, AI risks deepening rather than bridging educational inequalities. In this light, the future of education is not a technological question but a humanistic one: how can technology serve human flourishing rather than efficiency alone?

Across millennia, the impact of technology on education reveals a consistent pattern. Each major innovation, from the abacus and writing systems to the printing press, mimeograph, overhead projector, calculator, computer, internet, and AI, has expanded what learners can do while also redistributing power over knowledge. Technologies have helped standardize curricula, scale instruction, and broaden access, but they have also created new forms of exclusion, dependence, and surveillance (Eisenstein, 1979; Saettler, 2004). This history shows that tools are never neutral; they carry assumptions about whose knowledge matters, which skills are valued, and what purposes education should serve (Eisenstein, 1979; Saettler, 2004).

Artificial intelligence extends this pattern. AI systems not only store and convey information but also analyze, predict, and generate. In doing so, they bring into sharper focus long-standing questions that earlier technologies only hinted at: Who designs the systems that shape learning? Whose data trains them? How are errors, biases, and harms identified and addressed? Just as the printing press demanded new forms of critical reading, and the internet demanded new forms of digital and information literacy, AI demands algorithmic literacy, data ethics, and human–machine collaboration skills (UNESCO, 2024).

History also offers grounds for cautious optimism. Past technological moral panics have often given way to more balanced understandings once educators have had time to adapt pedagogy, assessment, and policy. Calculators did not destroy mathematical thinking; they shifted emphasis toward conceptual understanding and problem-solving. Similarly, AI need not undermine human agency if it is framed as augmenting rather than replacing intelligence and if teachers retain authority over how it is used in classrooms (Holmes et al., 2019; Selwyn, 2019).

Technology's role in education is a human choice, not a technical inevitability. The same tools that can deepen inequality can also be used to close gaps if questions of access, training, governance, and ethics are addressed directly. As the chapters that follow turn to AI in teacher preparation, instruction, assessment, and policy, the historical perspective offered here remains essential: the central question is not whether technology will shape education, but how it will do so and in whose interests.

With this historical frame in place, Chapter 2 turns from long patterns to present realities by mapping where AI is already embedded in schooling and what it is doing to instruction, assessment, and decision-making. Rather than treating AI as a single tool, the next chapter clarifies the roles AI now plays across classrooms and systems, and it introduces the guardrails educators need so human judgment and equity remain central.

Chapter 2:
The Role of AI in Modern Education

Artificial intelligence (AI) now shapes daily classroom work and district operations. It sits inside lesson-planning tools, literacy platforms, translation and accessibility supports, tutoring assistants, analytics dashboards, and the communication systems schools use every day. Students use AI to generate ideas, revise writing, and access supports; teachers use it to draft materials, analyze patterns, and reduce routine workload; leaders use it to surface early warning signals and organize interventions. This chapter clarifies what AI is doing in modern education, where it genuinely helps, and what guardrails are needed so human judgment stays central.

AI's expanding influence in education stems from both educational necessity and technological advancement. Educators face rising expectations and increasingly diverse student populations with a wide range of academic abilities and linguistic backgrounds. At the same time, the volume of data requiring analysis has grown substantially, while demands for individualized instruction, effective communication, timely intervention, and careful documentation have intensified.

In this context, AI offers scalable forms of support that traditional approaches often cannot provide on their own. Yet its integration into education also introduces serious challenges. It raises ethical, pedagogical, and societal concerns related to fairness, privacy, bias, academic integrity, and the possible erosion of essential human capacities. Although AI can strengthen instructional effectiveness, its misuse may also weaken learning. Educators, administrators, and policymakers must therefore balance innovation with careful oversight.

Chapter 1 traced how educational tools repeatedly reshape teaching, learning, and what schools choose to measure. Rather than revisiting that full history here, this chapter focuses on what educators and administrators are encountering right now: AI embedded in everyday platforms, accelerating planning, feedback, translation, analytics, and communication, and raising new questions because it can generate, interpret, and recommend, not just deliver information. In other words, Chapter 2 serves as the book's present-day map of AI's roles in education and the practical guardrails needed to use it responsibly.

AI's Distinct Strengths: From Information Storage to Creative Collaboration

Artificial intelligence marks a major shift in educational technology, one that moves far beyond the tools educators have traditionally relied upon to store or display information. Throughout the history of schooling, new technologies have expanded teachers' access to resources and streamlined administrative tasks. Yet these tools, despite their usefulness, never altered the cognitive foundation of teaching and learning. They improved efficiency but did not engage in the intellectual work themselves. In contrast, AI introduces capabilities that blur the boundaries between human cognition and technological assistance. Rather than serving only as repositories of information, contemporary AI systems can analyze, interpret, generate, and even co-create with learners and educators in dynamic ways.

This shift matters because AI now performs tasks that once depended almost entirely on human judgment. While a calculator can solve equations and a database can store facts, AI can read and summarize complex passages, evaluate student responses for meaning, generate original text, adapt explanations in real-time, predict learning patterns, and sustain conversational dialogue that responds to the learner's needs. In essence, AI operates as a cognitive collaborator capable of augmenting human thinking rather than simply accelerating mechanical or clerical tasks.

When applied to classroom practice, these capabilities expand what educators can ask students to do and what students can reasonably accomplish. For instance, AI can transform a single passage into multiple versions calibrated to different reading levels, allowing teachers to provide accessible materials without spending hours rewriting text. It can analyze student writing to identify strengths and misconceptions, offering suggestions for revision that mirror the targeted feedback a teacher might give during a conference. Beyond language-based tasks, AI can walk a student step-by-step through a math problem, offering hints and explanations as the student progresses, much like an individualized tutor who adapts to the learner's evolving understanding.

AI also reshapes instructional planning. With increasing accuracy, AI systems can draft lesson plans aligned to state standards, propose activities differentiated for varied readiness levels, and generate practice tasks that progressively increase in cognitive demand. Teachers, who often spend significant time crafting materials, designing assessments, and differentiating activities, can use AI to offload these time-intensive tasks and devote more energy to instructional decision-making, student relationships, and deeper pedagogical reflection.AI can function as a generative partner, producing examples, analogies, rubrics, and instructional scaffolds aligned with a teacher's goals.

Language learning and accessibility also benefit greatly from these AI capabilities. For instance, AI can translate directions, simplify academic vocabulary, or create bilingual glossaries that help multilingual learners access grade-level content. Students with disabilities can use AI to convert text to speech, reorganize dense passages into manageable summaries, or transform visual information into auditory formats. In these ways, AI extends the reach of inclusive practices and brings the principles of Universal Design for Learning (UDL) within practical reach for busy teachers.

The rise of learning analytics has long suggested that patterns in learner behavior can be used to inform instructional decisions and identify unmet needs more effectively (Siemens & Long, 2011). One of AI's most useful strengths is its ability to detect patterns across student data that once required extensive analysis or remained invisible altogether. AI can identify recurring errors in math assignments across a class, highlight reading skills that consistently challenge certain groups of students or alert a teacher to a student who is beginning to disengage. These diagnostic capacities allow educators to intervene earlier and with sharper precision. Rather than relying solely on periodic assessments, teachers can monitor learning as a continuous, unfolding process. Data-rich learning environments require careful attention to how learners and teachers interpret data, what kinds of reasoning dashboards promote, and how such information is used to guide instruction (Lee & Wilkerson, 2018).

AI can support creative and higher-order thinking by serving as a co-creator. Students can brainstorm research ideas with an AI partner, explore multiple angles on a social studies simulation, or generate scientific hypotheses to test in classroom experiments. Although artificial intelligence does not supplant human creativity, it enhances the learner's ability to investigate concepts by providing prompts, refinements, and diverse viewpoints that contribute to a more robust creative process. This co-creative dynamic encourages students to take intellectual risks and expand their thinking within guided structures. These capabilities are reshaping how teachers plan instruction, differentiate support, and respond to student learning. AI can expand access and responsiveness, but its educational value still depends on thoughtful use within sound teaching practice.

Factors Contributing to the Rapid Advancement of AI in Education

The rapid rise of artificial intelligence (AI) in education did not occur in isolation; rather, it emerged from a convergence of social, technological, and institutional pressures that reshaped the daily realities of teaching and learning. Over the past decade, educators have navigated increasing demands while also contending with unprecedented disruptions to schooling. These pressures created fertile ground for the swift adoption of AI tools, which promised not only efficiency but a fundamental rethinking of instructional support.

One of the most significant forces accelerating AI adoption has been the growing strain on teacher workload. Modern educators operate in an environment characterized by escalating expectations and limited resources. In addition to their core responsibilities of instruction and classroom management, teachers are required to plan differentiated lessons, design assessments, analyze student data, communicate with families, document interventions, provide feedback, and comply with numerous administrative mandates. Many teachers describe their workload as unsustainable, with planning and grading alone consuming hours beyond the school day. AI has emerged as a practical support, helping automate or streamline tasks that have traditionally placed heavy demands on educators. The promise of reduced cognitive and administrative load made AI immediately attractive to teachers who felt stretched thin.

At the same time, classrooms are now more diverse than ever, requiring adaptable and individualized teaching methods. Multilingual learners require linguistic scaffolds and accessible materials; students with disabilities benefit from adaptive supports that respond to their cognitive and physical needs; advanced learners require enrichment opportunities; and students whose learning was disrupted by the COVID-19 pandemic or other circumstances often need targeted interventions to rebuild foundational skills. While educators have long embraced the principle of differentiation, the practical reality of meeting the needs of every learner has always been complex. AI tools, with their ability to rewrite texts at varied reading levels, generate multiple examples, provide individualized feedback, and translate or modify content instantly, offer a new way to operationalize personalized learning at scale. AI enables teachers to customize lessons efficiently despite time and workload limitations.

The rapid adoption of AI was further enabled by the digital platforms schools had already put in place. Over the past decade, and particularly accelerated by pandemic-era remote learning, districts invested in one-to-one device initiatives, remote internet access, cloud-based platforms, digital assessments, and online curriculum tools. These infrastructures created an environment in which AI could be integrated with minimal logistical friction. Teachers and students were already accustomed to submitting assignments online, interacting through learning management systems, and using digital tools as part of daily instruction. The adoption of AI felt less like a technological revolution and more like a natural extension of existing practices, seamlessly fitting into digital workflows that educators were already navigating.

The rapid maturation of AI technology itself also played a crucial role. Contemporary large language models, adaptive learning systems, and generative tools reached a level of sophistication that made them both accessible and affordable for schools (Holmes et al., 2019). Previous generations of educational AI tools were often limited in accuracy, expensive to implement, or difficult for teachers to use. Fortunately, newer iterations of AI systems demonstrated unprecedented fluency in natural language, impressive adaptability to user input, and user-friendly interfaces that required little technical training. This combination of power and usability led many educators to experiment with AI tools and quickly integrate them into their instructional routines.

Finally, broader societal and economic shifts reinforced the necessity of preparing students for a workforce increasingly shaped by automation, data-driven decision-making, and collaboration with AI. Global organizations, including UNESCO (2024), have emphasized the importance of equipping learners with the skills needed to navigate a world in which AI systems are embedded across industries. Educators recognized that students must not only be able to use AI tools but also understand how to think critically alongside them, evaluating outputs, making informed decisions, and engaging in ethical reasoning. As a result, integrating AI into teaching and assessment became part of a broader strategy to prepare students for future professional environments that will demand these competencies. These conditions have made AI both appealing and difficult for schools to ignore. Its rapid adoption reflects not only technological progress but also the practical pressures facing contemporary education.

AI as an Instructional Partner

Artificial intelligence increasingly functions as an instructional partner, reshaping how teachers design learning experiences and how students engage with academic tasks. AI augments teachers' ability to address diverse classroom needs, rather than replacing them. In daily practice, teachers can turn to AI to alleviate burdensome tasks that once consumed significant portions of their professional time. Tasks such as generating differentiated texts, designing multiple versions of a reading passage, constructing rubrics aligned to standards, or drafting communications to students and parents can now be completed in minutes. These efficiencies enable educators to focus their expertise on instructional decision-making, fostering meaningful relationships, and delivering adaptive instruction: domains in which human discernment and professional insight continue to be essential.

The instructional benefits of AI extend deeply into planning and assessment. Teachers can quickly produce small-group lesson plans tailored to emerging needs they observe during instruction or through student data. When confronted with large amounts of formative assessment information, AI enables educators to analyze patterns rapidly, flag misconceptions, and identify students in need of specific interventions. Teachers who once spent hours reviewing exit tickets, writing feedback, or curating targeted practice materials can now generate these resources with the support of AI, ensuring that instructional actions are timely and data-informed. With AI taking on routine cognitive labor, teachers can engage more fully in collaborative planning, deeper analysis of student thinking, and the creative work of designing meaningful learning experiences in a more efficient manner.

Students encounter AI differently but no less impactfully. Many experience AI not as a planning tool but as an interactive support that accompanies them throughout the learning process. Tools powered by AI offer immediate feedback on practice problems, helping students understand missteps in real-time rather than waiting days or even weeks for graded assignments and tests to be returned. For multilingual learners, AI systems can translate content, simplify complex vocabulary, or provide bilingual explanations, greatly expanding access to grade-level material. Students who struggle with reading can listen to text read aloud, see vocabulary defined and contextualized, and receive guiding questions that foster comprehension and engagement. These supports make learning more accessible and help students maintain momentum even when they encounter difficulty or when the teacher is not present.

AI also plays a significant role in writing development. Students can receive feedback on organization, clarity, grammar, and coherence, enabling them to revise drafts with greater independence. These tools encourage iterative writing, which helps students build stamina and confidence. In research tasks, AI can support students by summarizing sources, suggesting areas of focus, or helping them structure inquiries. Importantly, when used appropriately, AI should not do students' intellectual work; it should help them understand complex tasks, manage information, and develop their ideas.

For learners who express themselves better through modalities other than writing, AI offers multimodal avenues for communication and creativity. Students may generate visual representations of concepts, convert text into audio explanations, or build interactive models that demonstrate their understanding. These tools allow students to demonstrate understanding in formats aligned with their strengths and preferences, thereby broadening the range of acceptable academic expression. These developments show that AI is becoming a practical support for both teaching and learning. Its effectiveness, however, still depends on teachers' ability to guide its use with professional judgment and clear instructional purpose.

Opportunities Presented by AI in Modern Education

1. **Personalized Learning at Scale**: The modern classroom is characterized by extraordinary diversity in student readiness, learning profiles, and academic needs. In a typical middle school environment, for example, teachers may encounter students reading several grade levels apart, learners who possess strong background knowledge alongside those who are encountering material for the first time, and a wide range of motivational and cognitive differences. Although differentiation has long been championed as a best practice, many educators acknowledge that providing multiple pathways for every lesson often feels impossible given the constraints of class size and instructional time. With twenty-five to thirty-five

students in a fifty-minute class period, the scope of individualized support that a teacher can reasonably provide is inherently limited.

2. **Enhanced Accessibility and Equity**: Classrooms today reflect a rich tapestry of languages, cultures, and learning needs. In regions such as Southern California, it is common for high school teachers to instruct students who speak multiple home languages, each requiring different levels of linguistic support. Before AI-based translation and accessibility tools became available, teachers often relied on peer translators, visual gestures, or limited bilingual dictionaries to communicate complex ideas. These strategies, while helpful, were insufficient for providing equitable access to academic instruction.

3. **Reduced Teacher Workload and Professional Sustainability**: Educators overwhelmingly enter the profession because they value teaching, intellectual engagement, and student relationships. Yet national surveys consistently show that teachers spend a disproportionate portion of their time on administrative work: planning, grading, documenting interventions, communicating with families, managing digital platforms, and preparing compliance reports. These responsibilities not only encroach upon planning and instructional time but also contribute substantively to burnout and attrition across the profession.

4. **Expanded Creative Possibilities**: The creative potential of AI has opened new pathways for expression across academic disciplines. In English classes, for example, students can use AI tools to brainstorm ideas, explore narrative possibilities, generate outlines, or refine sentence structures. Rather than diminishing creativity, AI is a catalyst that helps students who

struggle with idea generation, or organization find a starting point for their work. In art and design courses, AI coding tools and image generators allow students to conceptualize visual compositions, experiment with textures, and explore stylistic variations. Meanwhile, in disciplines such as science, technology, engineering, and mathematics (STEM), AI-powered simulations enable students to model scientific phenomena, conduct virtual experiments, or approach engineering challenges with tools that were previously limited by classroom resources or equipment availability.

5. **Improved Decision-Making through Data Insights**: Schools generate vast amounts of data, yet educators frequently lack the time, tools, or support needed to interpret it. Even for seasoned educators, navigating the complex maze created by grades, behavior logs, attendance records, benchmark assessments, and progress monitoring tools can be overwhelming. As a result, early warning signs of academic decline or disengagement often go unnoticed, not because educators are inattentive, but because the volume of information exceeds human ability for analysis.

AI can automate or accelerate many of the tasks that contribute to this workload. Teachers can use AI to draft lesson plans aligned with state standards, construct rubrics, design assessments, create differentiated materials, and summarize performance data. AI can also assist with drafting emails, parent newsletters, professional reflections, or meeting summaries. It can help organize behavioral and intervention documentation, streamlining processes that once required considerable manual effort. Many educators who are comfortable using AI report that they gain back valuable time, allowing them to reinvest in instructional preparation, relational work with students, and their own well-being. AI can contribute to the long-term sustainability of the teaching profession.

Challenges and Risks of AI in Contemporary Education

- **Algorithmic Bias and Fairness Concerns**: While AI offers enormous potential for equity, it can also amplify inequities if not carefully implemented. AI models learn from existing data that may reflect historic biases, stereotypes, or unequal treatment. When biased datasets influence an AI tool, the resulting recommendations can be inaccurate or unfair. Examples include predictive models that disproportionately flag students from marginalized backgrounds as at risk, writing analysis tools that penalize dialects such as African American Vernacular English, automated scoring systems that misinterpret unconventional writing structures, and recommendation algorithms that steer some students away from advanced coursework.

  These risks require continuous monitoring and a commitment to fairness. Educators must ask: Whose data shaped this tool? Who stands to benefit and who might be harmed? Bias is not a flaw in AI alone; it mirrors inequities in society. Addressing this challenge demands transparency, ongoing evaluation, and involvement of diverse stakeholders.

- **Privacy, Data Security, and Ethical Use of Student Information**: AI systems often require substantial data inputs, such as writing samples, performance metrics, engagement records, and personal identifiers. This raises concerns about how schools collect, store, and share student information. If not carefully managed, AI adoption and use may lead to inadvertent privacy violations or the misuse of sensitive data. Common concerns include how long student data is stored, whether vendors use data to train commercial models, whether parents and students are fully informed, what safeguards protect against breaches, and whether systems comply with the

Family Educational Rights and Privacy Act (FERPA) and state privacy laws.

One district in Texas suspended the use of an AI tool after learning that the vendor stored student writing samples long term without clear consent (Cerullo, 2023). Such examples highlight the need for rigorous data governance policies, transparent communication, and careful contract review. Protecting student privacy must remain a top priority in AI integration. Trust is foundational to education, and any breach of trust threatens the legitimacy of AI adoption.

- **Overreliance on AI and the Erosion of Critical Thinking**: AI tools can generate answers quickly, sometimes too quickly. Students who discover that AI can summarize chapters, solve math problems, or draft essays may be tempted to skip the learning process altogether. This risk is not theoretical; teachers increasingly report students turning to AI before attempting tasks themselves. Social media environments frequently expose students to influencers and others who misuse AI to attack followers or sell products.

Scholars have warned that overreliance on AI may weaken critical thinking, metacognition, perseverance, creativity, and deep comprehension if students use it to bypass rather than support learning (Holmes et al., 2019; Selwyn, 2019). To address this challenge, educators must fully understand the benefits and risks associated with AI themselves and then teach students how to use AI strategically and ethically, just as they would with any other tool. AI should always support learning, not circumvent it. Student instruction must include explicit discussions about when AI is appropriate, how to interrogate AI-generated content, and how to maintain academic honesty.

- **Academic Integrity in the Age of AI**: AI fundamentally changes how students produce work. An AI-generated paragraph can appear flawless, making it difficult to determine who authored what. Traditional plagiarism detectors are increasingly unreliable in detecting AI-generated text because they struggle with accuracy, produce false positives and negatives, and often demonstrate bias against non-native English speakers. This shift requires schools to rethink academic integrity policies from the ground up. Key policy considerations should include clear definitions of acceptable AI use, expectations for citation or disclosure, assignments that prioritize process over product, oral defenses or in-class tasks, and reflection-based assessments. Integrity becomes less about policing misconduct and more about guiding students through responsible, transparent collaboration with AI.

- **Resistance, Fear, and Shifting Teacher Identity**: AI challenges long-held beliefs about what teaching is and what teaching should be. Some educators fear being replaced. Others worry that students will value AI more than human instruction. Many teachers feel unprepared or intimidated by the rapid pace of technological change that has occurred in the last decade. These concerns are valid and deserve serious attention by school administrators and school boards. Recent research similarly indicates that teachers often approach AI with a mixture of interest, uncertainty, and concern about how it may affect professional practice and student learning (Uygun, 2024). Teachers must be supported through professional development, peer collaboration, coaching, thoughtful implementation, and reassurance that human relationships remain irreplaceable. AI alters the tasks of teaching but not its fundamental essence. The empathy, mentorship, and emotional insight teachers provide cannot be replicated by any machine or algorithm.

- **Implementation Gaps and Unequal Access**: AI's benefits are unevenly distributed. Schools with robust funding can experiment with AI tools, provide training, and infrastructure upgrades. Under-resourced schools may struggle with limited devices, unreliable internet, outdated hardware, insufficient staffing, and a lack of training. This digital divide risks widening achievement gaps across districts and states rather than closing them. For AI to enhance equity, policymakers and leaders must ensure that all schools, not just affluent ones, can implement AI responsibly.

Framework for Ethical and Effective AI Integration

AI is most effective when educators remain the final decision makers. The Human-in-the-Loop model ensures that technology supports, not dictates, teaching and learning (Wu, 2022; Benedikt et al., 2020). This model recognizes that teachers must interpret, approve, and monitor AI-generated suggestions, and that students must be taught to reflect on and verify AI-supported feedback rather than accept it uncritically. To keep human judgment central, schools can organize responsible AI implementation around the following six pillars:

- **Transparency:** Schools should be able to explain, in plain language, when AI is being used, for what purpose, what data it draws from, and what it will and will not do well. Transparency also includes disclosure norms for students, e.g., when AI assistance must be acknowledged, and clear communication with families about how AI-supported tools interact with student work. In practice, transparency looks like visible tool lists, parent-friendly explanations, and educator guidance that treats AI output as "suggestions with limits," not hidden authority.

- **Accountability:** AI can inform decisions, but it cannot be responsible for them; responsibility must remain with educators, schools, and districts. Accountability requires clear decision rights that specify who may adopt tools and how they may be used, defined review processes, and mechanisms for addressing errors, harm, or disputes. It also includes meaningful recourse. Students and families need a pathway to question or appeal AI-influenced judgments, especially in high-stakes areas such as grading, placement, discipline, or intervention.

- **Equity:** Equity means AI is implemented in ways that expand opportunity rather than reproduce historic patterns of bias, surveillance, and unequal access. Schools should examine how tools perform for multilingual learners, students with disabilities, and students whose language or culture differs from dominant training data, and they should monitor for disparate outcomes over time. Equity also includes infrastructure and support: if some schools lack devices, bandwidth, or training, AI can widen gaps unless systems invest deliberately to close them.

- **Privacy:** Privacy protects student dignity by limiting what data is collected, how long it is kept, who can access it, and for what else it can be used. In practical terms, privacy includes data minimization, clear retention and deletion timelines, strong security controls, FERPA-aligned practices, and contract language that restricts vendor secondary use, including model training, unless explicitly permitted. Privacy is also relational: families are more likely to trust AI-supported learning when schools communicate data practices clearly and avoid normalizing surveillance as a default.

- **Governance:** Governance is the system-level structure that makes responsible use sustainable: who approves tools, how pilots are evaluated, what documentation is required, and how policies are updated as technology changes. Good governance prevents "pilot sprawl" by setting consistent expectations for procurement, staff training, monitoring, and incident response. It also creates routines, e.g., semester reviews, so schools can audit for bias, address emerging risks, and discontinue tools that do not meet ethical or instructional standards.

- **Human Judgment:** Human judgment is the safeguard that keeps education ethical and context-sensitive. Teachers and leaders must verify AI outputs, interpret recommendations with knowledge of student context, and decide when not to use AI, especially in high-stakes moments where misclassification can cause lasting harm. Human judgment also protects professional autonomy and student agency: AI may accelerate drafting, feedback, or analysis, but it should never replace the teacher's responsibility to decide what is developmentally appropriate, culturally responsive, and fair.

These pillars are not a compliance checklist; they are a practical way to translate educational values into everyday decisions about tools, data, instruction, assessment, and communication. Together, they clarify what must be visible and explainable (transparency), who remains responsible (accountability), how harms and disparities are prevented (equity and privacy), how systems sustain consistent practice (governance), and where professional discernment must remain decisive (human judgment).

Building AI Literacy for Educators

As artificial intelligence (AI) becomes increasingly woven into the fabric of educational practice, the need for robust AI literacy among educators has grown more urgent. While teachers are not required to become computer scientists or AI engineers, they do require a foundational understanding of how AI systems function, how to interact with them responsibly, and how to integrate them into instruction. This form of literacy empowers teachers to make informed decisions, to safeguard students, and to preserve the integrity of the learning process.

AI literacy for educators begins with a conceptual understanding of how AI tools operate. Teachers need to understand that AI systems produce results from learned data patterns and do not possess human-like thinking or knowledge. This conceptual grounding helps educators recognize the limitations of AI, particularly the potential for inaccurate or fabricated information. Knowing that AI can hallucinate facts, misinterpret complex student responses, or offer overly confident explanations is essential for preventing misuse. Educators must be able to critically evaluate AI outputs, cross-check information, and discern when an AI tool is offering reliable guidance versus when it may be producing errors or misinformation.

It is just as crucial for teachers to recognize and address any bias found in AI-generated content. Since AI models are trained on large datasets that may include societal biases, stereotypes, or inequitable representations of groups, teachers must be alert to the ways bias can manifest in AI suggestions, translations, or examples. Developing this awareness ensures that AI does not inadvertently reinforce harmful assumptions or inequities in instructional materials. Educators must be prepared to question outputs that seem skewed or misaligned with inclusive teaching practices, and they should be equipped with strategies to correct or contextualize biased content.

Crafting effective prompts, the process often referred to as "prompt engineering," is becoming a foundational skill for educators using generative AI systems (Montalvo, 2025). Teachers must learn how to formulate clear, specific, and purposeful prompts to guide AI tools toward producing appropriate instructional materials. A well-crafted prompt can help teachers generate differentiated texts, design practice problems, or create lesson ideas that support students' needs. Conversely, vague or poorly directional prompts can lead to outputs that are confusing, developmentally inappropriate, or misaligned with learning goals. Understanding how to guide AI tools through iterative prompting, refining inputs until desired results are achieved, is a practical competency educators must develop.

Another dimension of AI literacy involves the thoughtful integration of AI into lesson design. Teachers need to determine when AI enhances learning and when it is a distraction or a shortcut that undermines deeper comprehension. Integrating AI responsibly requires a nuanced approach: teachers must align AI use with standards, learning targets, and pedagogical goals. They must decide when AI should assist in generating materials or offering feedback, and when students should engage in independent problem-solving without technological support. This discernment preserves the integrity of the learning process while leveraging AI's strengths where appropriate.

Evaluating AI-supported student work presents a new challenge that educators must navigate. As students increasingly use AI for drafting, revising, or brainstorming content, teachers need clear frameworks for distinguishing student-generated thinking from AI-generated content. This includes developing rubrics that account for AI assistance, asking students to document their use of AI, or incorporating reflective components that reveal the learner's understanding. Teachers must learn to assess not only the final product but also the processes students use, thereby ensuring that AI enhances learning rather than replacing the cognitive work students are expected to perform.

Comprehensive professional learning is essential to support the development of this AI literacy. One-time workshops are insufficient for a technology that evolves rapidly and poses complex academic, ethical, and instructional challenges. Professional development must be ongoing, collaborative, and grounded in real classroom experiences. Teachers benefit from opportunities to experiment with AI tools, reflect on successes and challenges with colleagues, examine case studies, and receive coaching that connects AI integration directly to curriculum and assessment goals. Such sustained learning fosters confidence, competence, and responsible practice.

As AI becomes a regular part of students' daily lives, educators must also intentionally build AI literacy among their learners. Students need explicit instruction in how to interact with AI ethically, responsibly, and effectively. This includes understanding when AI use is appropriate, such as brainstorming or clarifying directions, and when it is inappropriate, such as completing assignments that require original thought or independent analysis. Students must learn to evaluate AI-generated content critically, recognizing that the tool may provide inaccurate, biased, or incomplete information. Teaching students to verify sources, question outputs, and apply their own reasoning is essential to developing informed digital citizens.

Students must learn and adapt academic integrity practices to the AI era. This means knowing how to cite AI contributions, understanding the boundaries between AI support and academic dishonesty, and appreciating the value of doing their own cognitive work. By learning how to use AI as a partner rather than a substitute for learning, students cultivate habits of mind that strengthen their critical thinking and prepare them for future challenges.

AI's value in schools will not be determined by how impressive the tools are, but by the professional judgment and local decisions that shape how they are used. For teachers, the practical question is when AI strengthens learning and when it becomes a shortcut that weakens it. For school and district leaders, the practical question is how to select tools, protect students, and build shared expectations so AI use is consistent, ethical, and equitable. The next chapter turns to what this means for teacher preparation, because the speed of AI adoption has outpaced what most educators were ever trained to supervise.

Building AI literacy, both for educators and students, is less about mastering a technology and more about shaping the human–technology partnership that defines contemporary teaching and learning. As educators adapt to utilizing the capabilities of AI while preserving their professional judgment, and as students approach AI with diligence and ethical awareness, educational institutions are increasingly prepared to navigate an environment where AI serves not only as a tool but as a significant factor shaping communication, innovation, and the development of knowledge.

Taken together, these opportunities and risks show why AI cannot be treated as a plugin feature or a purely technical upgrade. When a system can generate feedback, recommend interventions, and shape how students demonstrate learning, educators need shared expectations for transparency, privacy, equity, and human oversight. Chapter 3 turns to what this means for teacher education by asking how preparation programs can equip new teachers to supervise AI responsibly, not just use it.

Part II:
Equipping Educators for
the Age of AI

Chapter 3:
The Changing Role of Technology
in Teacher Education

Teacher education has long evolved alongside the tools and instructional possibilities of its time. From the 19th-century normal schools that trained teachers in chalkboard technique and blackboard penmanship to modern colleges of education that include digital pedagogy, learning analytics, and virtual simulations, teacher preparation has evolved in step with broader technological and social change. Yet no era has reshaped teacher education as rapidly as the rise of artificial intelligence. Unlike earlier educational tools, AI reshapes not only the materials teachers use but the cognitive and instructional foundations of teaching. It can generate lesson plans, model classroom interactions, analyze student performance, and provide automated feedback, placing technology closer to the work of professional judgment than most prior innovations. This shift raises new expectations for teacher expertise: candidates must learn to supervise AI outputs, recognize bias and error, protect student privacy, and decide when human judgment must override automation. Teacher preparation must therefore move beyond "how to use tools" toward "how to govern tools" ethically and pedagogically within real classroom contexts.

The preceding two chapters established two ideas that matter for teacher education: educational technologies reshape practice, and AI is different because it can generate, interpret, and recommend, not just deliver content. This chapter turns that lens toward teacher preparation. Rather than re-telling the broader history of schooling technology, it focuses on how preparation programs have historically responded to major shifts, from the chalkboard era through the rise of computers, and why AI now requires new forms of professional judgment, ethics, and supervision embedded across coursework and clinical practice.

To make that case, this chapter traces how teacher education responded to earlier technological eras: the chalkboard and recitation model, the mechanical and analog expansion of reproducible materials and media, the computer and internet push for digital literacy, and the interactive and mobile shift toward blended and data-informed instruction. Across these eras, programs often focused on operational competence, emphasizing how to use the tool, while underestimating how technology changes what counts as good teaching, how learning is assessed, and whose experiences are centered. AI is the clearest test of that pattern because it can generate, interpret, and recommend in ways that affect feedback, assessment, and opportunity. The historical arc that follows is meant to help readers see what is continuous (implementation cycles, uneven adoption, moral panic) and what is new (algorithmic mediation of judgment, privacy risk, and equity impact).

The Chalkboard Era (1800s–1940s)

The chalkboard, first widely adopted in the early 1800s, was the earliest classroom technology to fundamentally alter teaching practice. Prior to its introduction, teachers relied heavily on oral recitation and individual slates. With the chalkboard, whole-class instruction became possible. Teachers could model writing, arithmetic, diagrams, and demonstrations visible to an entire room of students. Normal schools, the predecessors of today's colleges of education, explicitly trained teachers in chalkboard handwriting, board organization and sequencing, diagrammatic instruction, and pacing techniques for large groups. Teacher candidates practiced board work as an essential skill, demonstrating mastery by delivering "model lessons" in front of peers. The chalkboard reinforced teacher authority, emphasizing clarity of presentation, structured lecture formats, and teacher-centered pedagogy. In these decades, technology amplified the teacher's voice, literally and figuratively, reinforcing cultural assumptions about the teacher as deliverer of knowledge.

The chalkboard influenced not only how teachers taught but also what teacher education valued. Presentation competence was central to teacher identity. Classroom management centered on visible, orderly recitation. Assessment focused on oral recall and board demonstrations, and materials such as chalk, erasers, and slates that shaped lesson design. The chalkboard also foreshadowed patterns visible in AI today: initial enthusiasm, uneven adoption, and debates about whether students should rely too much on teacher demonstrations instead of developing their own methods. In this period, technology remained an extension of teacher skill, not yet a partner in cognition.

The Mechanical and Analog Era (1940s–1980s)

The introduction of mechanical reproduction technologies in the mid-20th century dramatically changed both classroom practice and teacher preparation. The mimeograph, patented in the late 19th century but widely adopted by schools in the 1940s and 1950s, allowed teachers to duplicate worksheets and handouts rapidly (Saettler, 2004). Teacher candidates learned how to create stencils and templates, how to design reproducible materials, and how to manage classroom routines involving independent work. Reproducibility shifted classroom instruction from oral lessons to paper-based tasks, worksheets, and seatwork. It also changed the pace and scale of planning: teachers could prepare one set of materials for an entire class, reuse them across sections, and revise them between cycles of instruction without starting over. Over time, this supported a more standardized curriculum, expanded the use of common assessments, and strengthened the administrative logic of schooling, where what could be copied and distributed easily often became what was taught most often.

Introduced widely after World War II, overhead projectors provided teachers with reusable transparency sheets, the ability to annotate live in front of students, pre-made visual materials, and a new form of interactive demonstration. Teacher preparation programs incorporated instructional media coursework where candidates learned to design visual aids, integrate multimodal representations, and pace lessons using sequential transparencies. This technology enabled teachers to engage students visually while maintaining flexibility in lesson delivery, supporting both planned and spontaneous instruction. It also supported early formative assessment routines, as teachers could mark up student examples, model revisions, and make thinking visible in real time. Overhead projectors also encouraged collaborative learning, as students could interact with projected materials and participate in classroom discussions more actively.

The overhead projector was the precursor of today's interactive whiteboards, expanding visual learning and enabling more interactive teaching long before the digital era. Between the 1950s and 1970s, educational television and filmstrips introduced audiovisual learning. Teacher educators began discussing media literacy, student engagement through multimedia, the role of visuals in comprehension, and the risks of passive learning. Neil Postman (1985) famously warned that televised instruction risked turning classrooms into spaces of entertainment rather than inquiry, a critique echoed today regarding AI-generated content.

Calculators were introduced to classrooms gradually in the early to mid-1970s as affordable handheld models emerged, becoming commonplace in high school math by the late 1970s. After considerable debate on their potential impact on student cognition and math skills, the National Council of Teachers of Mathematics (NCTM) recommended their use at all levels by the early 1980s (Watters, 2015). During this era, teacher preparation expanded to include operating machines and calculators, integrating visual media, designing reproducible instructional materials, and discussing cognitive overload and attention. Technology supplemented teacher instruction, but teachers still had to orchestrate learning deliberately. The teacher was still the primary source of knowledge.

The Computer and Internet Era (1980s–2000s)

The introduction of personal computers in schools in the early 1980s marked the first major shift in teacher education since film and television. Teacher candidates were introduced to digital literacy, which was initially framed not as a pedagogical practice but as a technical skill. Colleges of education developed "Computer in Education" courses focused on basic keyboarding and typing, word processing, spreadsheets for grading, early educational software, and introductory programming (LOGO, BASIC). These courses treated computers as separate tools rather than as integrated pedagogical resources. Even so, the presence of computers fundamentally altered expectations of teacher preparation. Teachers were now expected to manage digital resources, store electronic gradebooks, and use computers for communication.

Seymour Papert's (1980) constructionism became particularly influential in teacher preparation by positioning computers as tools for discovery and creation, not just consumption. Instead of simply using computers for rote practice or information delivery, teacher candidates were encouraged to leverage technology for hands-on learning and problem-solving. Candidates used software to present material and encourage inquiry with simulations, modeling tools, and multimedia projects, fostering an environment where students could explore concepts, experiment with digital tools, and actively construct their own understanding. This approach shifted teacher preparation toward more student-centered, inquiry-driven practice and laid the foundation for later digital and AI-integrated instruction.

By the 1990s, computers had become an essential part of teacher preparation, fundamentally altering education. Teacher education programs expanded to include a wide range of digital skills, such as digital storytelling, multimedia presentations, early online research, internet-based classroom projects, email and online communication, and web literacy. Teacher candidates were expected not only to use technology but to integrate it into their lesson plans and instructional strategies. Integration, however, was not uniform; it often varied by discipline, grade level, and individual program. For example, social studies programs might emphasize online research and digital archives, while science education incorporated data collection and simulations.

Despite the push for integration, many faculty members struggled to model effective digital teaching. This resulted in inconsistencies between coursework and clinical practice, with some candidates receiving robust training and others encountering gaps in their preparation. The rapid pace of technological change also meant that teacher educators had to continually update their own skills and knowledge, further contributing to uneven implementation. These challenges underscored the need for clearer expectations and guidance to ensure that all teacher candidates were prepared for digital classrooms.

The National Educational Technology Standards (NETS), first introduced by the International Society for Technology in Education (ISTE) in 1998, provided a much-needed framework to help align teacher education with emerging technological expectations in schools (International Society for Technology in Education, 1998). These standards outlined clear competencies for teacher candidates, emphasizing the need to plan technology-enhanced lessons, evaluate and select appropriate digital tools, demonstrate ethical behavior and responsible use of technology, and teach students about online safety and digital citizenship. The NETS also encouraged teacher candidates to foster creativity, collaboration, and critical thinking through technology, moving beyond basic computer skills to more sophisticated pedagogical uses.

With the adoption of NETS, teacher education programs began to reshape their curricula, incorporating training on classroom internet use, digital assessment tools, and strategies for managing online student behaviors. Faculty received professional development to better model digital teaching, and programs increasingly emphasized the importance of preparing candidates for the ethical and equitable use of technology. These expectations laid the groundwork for the digital teaching competencies required today, setting the stage for ongoing innovation and adaptation as new technologies, such as learning management systems, interactive whiteboards, and AI-driven tools, became part of the classroom environment.

The World Wide Web transformed teacher preparation by ushering in a new era of connectivity and digital access. Teacher candidates needed to develop robust information literacy, including advanced online search strategies, critical evaluation of website credibility, integration of diverse online resources into lesson planning, and a thorough awareness of copyright, fair use, and digital ethics. The internet significantly reshaped teacher identity, shifting educators from primary sources of information to facilitators who help students navigate expanding digital knowledge, misinformation, and evolving resources.

In response, teacher educators updated methods courses to include online research projects, web quests, internet-based collaborative assignments, and digital communication tools. In this era, technology evolved from an instructional aid to a full-fledged pedagogical environment, requiring teachers to foster digital citizenship, model responsible online behavior, and adapt to the collaborative, interactive possibilities the Web provided.

The early 2000s brought widespread adoption of learning management systems (LMS) such as Blackboard, WebCT, and Moodle. Teacher candidates learned to upload assignments, participate in online discussions, manage digital grading, and design modules for blended or online learning. These systems foreshadowed modern platforms like Google Classroom and Canvas, which now offer even more integrated features for collaboration and assessment. LMS training equipped future teachers not only with technical skills but also with strategies for fostering student engagement and maintaining academic integrity in virtual settings. The ability to communicate with students and parents remotely became increasingly important, and these systems gave teachers new ways to track progress, provide feedback, and personalize instruction. Teachers' familiarity with LMS helped them adapt to remote communication and more flexible classroom models. This skill became crucial decades later during the COVID-19 pandemic, when schools had to quickly shift to online learning.

The rapid rise of computer-based instruction exposed inequalities in access, in that some schools lacked functioning computer labs. Rural areas struggled with internet connectivity, and families had unequal access to home computers, creating significant barriers to participation in digital learning. Teacher preparation programs began emphasizing digital equity, encouraging candidates to consider how socioeconomic factors, geographic location, and infrastructure shaped students' access to educational technology. Discussions expanded to include strategies for supporting students with limited resources, advocating for improved access, and designing lessons that could be delivered both online and offline. This focus on equitable access helped lay the groundwork for today's debates about data and AI literacy, which also require an equity-centered approach to ensure all learners benefit from technological advances regardless of their circumstances.

The Interactive and Mobile Era (2000s–2010s)

The introduction of interactive whiteboards (IWBs), often referred to by the popular brand name Smartboard, marked the next major technological shift in classrooms. Teacher candidates trained on interactive whiteboards learned to layer visual, auditory, and tactile elements, annotate digital content, create interactive simulations, and manipulate digital objects to model concepts. The interactive whiteboard served as a bridge between analog teaching and digital interactivity, allowing teachers to move fluidly between explanation, demonstration, and student participation. Teacher education programs redesigned media courses to teach IWB lesson design, meaning planning a lesson sequence that uses the board intentionally for interaction, with students manipulating content, embedded checks for understanding such as quick polls, sorting tasks, and annotation prompts, and smooth transitions between whole-group instruction and small-group or independent work. This technology challenged teacher candidates to think beyond static visuals, adopting multimodal strategies aligned with Universal Design for Learning (UDL) and making student interaction a deliberate part of the instructional plan.

The mid-2000s also brought laptops and netbooks into classrooms on an unprecedented scale. One-to-one device programs required teachers to manage digital workflows, incorporate online texts, use formative assessment apps, teach media literacy, and integrate collaborative tools like Google Docs. Teacher preparation programs revised coursework to include digital classroom management, device integration strategies, online collaboration norms, and safeguarding student data. Student teaching placements now involved managing classrooms where every student had a device, which represented a radical shift from chalk-and-board teaching.

Tablets like the iPad and Android devices further expanded instructional possibilities. Teacher candidates learned to use apps for leveled reading, creative expression, real-time feedback, targeted practice, and translation and accessibility. This helped usher in personalized learning, where software adapted tasks to students' needs. Over time, tablet ecosystems normalized the idea that instruction could be modular and data-informed, with dashboards, progress maps, and adaptive sequences shaping daily decisions about pacing and support.

Teacher education programs therefore began to emphasize app curation and evaluation, not only how to operate tools, but how to judge instructional quality, interpret the data a tool produces, and decide what actions to take next based on that evidence. These shifts also surfaced practical challenges that candidates had to learn to manage, including equitable access to devices, distraction and off-task behavior, and the risk that personalization becomes automated tracking rather than meaningful learning. This era paved the way for today's AI-driven adaptive tools, which function on the same principle but with far more sophistication.

Smartphones and tablets blurred the boundaries between formal and informal learning spaces, turning "mobile learning" from a niche strategy into an everyday reality for students and teachers. Because these devices travel with learners, they normalized micro-learning through watching short explanations, capturing images and notes, annotating readings, and continuing discussions outside of class, and they expanded multimodal production through audio, video, images, and collaborative documents as legitimate ways to demonstrate understanding. Teacher candidates explored mobile learning strategies such as QR code scavenger hunts, field-based data collection, augmented reality, video annotation tools, and on-the-go collaborative tasks.

At the same time, ubiquitous access created new classroom demands: attention management, notification distraction, and the need for clear norms about when devices support learning and when they undermine it. Mobile platforms also intensified privacy and data concerns, as apps track usage patterns, location data, and student interactions, often through vendor ecosystems that schools do not fully control. Finally, the mobile era sharpened equity questions: "bring your own device" policies and inconsistent connectivity meant that access, and the quality of participation, could vary dramatically by student and community. Teacher identity shifted again from in-class facilitator to designer of flexible, device-enabled learning experiences.

Beyond mobile hardware itself, the interactive and mobile era was defined by platform ecosystems that changed how teachers organized learning. Cloud-based collaboration tools made real-time co-authoring normal, shifting student work from isolated drafts to shared documents with revision histories, comments, and ongoing peer feedback. In many classrooms, this also changed what teachers assessed: process evidence, including revision, participation, and collaboration, became more visible, while "final products" became less distinct. Teacher preparation programs increasingly had to address the pedagogy of collaboration by teaching candidates how to design group work, teach feedback norms, and manage shared accountability, because the tools made collaboration easy but not automatically productive. These shifts also foreshadowed a central AI-era challenge: when systems automate parts of drafting and revising, teachers must decide what counts as authentic evidence of learning.

The period also ushered in participatory media: social platforms, short-form video, and user-generated content became major sources of information for students and therefore major variables in classroom attention, belief formation, and academic integrity. Video platforms supported "flipped" approaches in which students encountered content outside class and used class time for discussion, problem solving, and application, but this model depended on equitable access and careful design. At the same time, the spread of social media made digital citizenship more than an abstract concept. Teacher candidates needed practical strategies for teaching source evaluation, handling misinformation, setting boundaries around communication and online behavior, and supporting student well-being in always-connected environments (Selwyn, 2019). In many ways, this era shifted teacher preparation from technology integration toward media literacy, ethical participation, and the management of learning ecosystems that extended far beyond the classroom walls.

Massive Open Online Courses (MOOCs) and the open education movement changed how teachers accessed professional learning. Teacher candidates increasingly encountered free online university courses, open-access textbooks, and global discussion forums that made preservice learning easier and more efficient. Faculty also drew from open education resources (OER) for lesson examples and materials. MOOCs further prepared teachers for digital learning environments by introducing them to large-scale online instruction, asynchronous learning facilitation, and multimedia lecture design, which foreshadowed the remote instruction boom that would define the 2020 pandemic response. This period also normalized micro-credentialing and self-directed professional learning pathways, signaling that teacher learning could be continuous, modular, and networked rather than confined to formal coursework.

The late 2000s and early 2010s marked the rise of learning analytics dashboards. Teacher candidates were introduced to digital gradebooks, web-based formative assessment tools, data visualization, skill-progress mapping, and early predictive analytics. Teacher preparation programs began to incorporate data literacy, training candidates to interpret real-time student data to adjust instruction. This created the foundation for understanding AI-supported analytics in modern classrooms.

The AI Era (2010s–2020s)

The emergence of AI marked a significant shift in teacher education. Unlike earlier technologies, which primarily enhanced efficiency, visualization, or communication, AI transforms cognition, analysis, and creation. It is both a content generator and an interpreter, modeling complex reasoning alongside teachers. Its integration demands new forms of expertise, new ethical frameworks, and reimagined teacher identities. In teacher education, AI's defining feature is its ability to perform tasks that previously required human judgment or labor. AI tools can generate lesson plans, draft communication to parents, design assessment items, provide feedback on student writing, identify misconceptions, analyze data trends, and model classroom interactions.

For teacher candidates, AI acts as a co-designer that accelerates planning and expands pedagogical options. For example, a candidate can input a standard, such as "interpret figurative language" or "multiplying fractions," and receive sample lesson sequences, differentiated materials, formative assessment items, and targeted feedback prompts. This changes the cognitive environment of teacher preparation. Candidates must critically evaluate AI suggestions, recognizing that convenience does not ensure accuracy or pedagogical soundness. Teacher educators must emphasize that AI-generated content is a starting point, not a finished product. AI amplifies teacher creativity but does not substitute for instructional decision-making.

AI-driven writing analysis tools, adaptive math systems, and intelligent tutoring programs allow teacher candidates to observe automated feedback loops, learning pathways, data dashboards, misclassification risks, and differential impacts on diverse learners. Candidates learn to interpret AI insights while maintaining professional responsibility for grading and assessment decisions. They must understand when AI is helpful, when it is misleading, and when human judgment must override automation. Assessment literacy now includes AI literacy, particularly around bias, hallucinations, and the limitations of predictive modeling.

One of the most important applications of artificial intelligence in teacher education is its use in immersive simulations. Teacher candidates can now practice classroom management, parent-teacher conferences, culturally responsive dialogue, responding to misconceptions, facilitating academic discussions, and delivering feedback. AI-powered avatars and virtual environments provide immediate feedback, repeated rehearsal, exposure to diverse student behaviors, and safe spaces to navigate high-stakes scenarios. Simulated classrooms, such as Mursion, an AI-powered upskilling platform, or AI-driven role-play systems, reduce the "practice shock" historically experienced by new teachers entering real classrooms. The integration of simulation into teacher preparation parallels flight simulators in aviation: teachers rehearse complex tasks before performing them with real students.

Ethical, Pedagogical, and Equity Implications of AI

AI introduces unprecedented ethical responsibilities into the teaching profession because it can influence how students are evaluated, supported, and understood. One immediate concern is algorithmic bias. Candidates must learn how AI systems can reproduce or amplify inequities in writing feedback, behavior analysis, predictive risk models, and automated grading, especially when training data reflects historic disparities. Teacher preparation should therefore treat bias detection as a practical skill, not an abstract principle, by giving candidates repeated opportunities to critique AI outputs, compare them with human judgment, and ask who benefits and who is harmed.

A second responsibility is data governance and privacy. Because many AI tools depend on student writing, performance data, and engagement signals, candidates must understand how FERPA compliance, ethical data use, and vendor transparency should shape decisions about adoption and classroom use. This includes practical questions that arise in everyday teaching: what data can be entered into a tool, how long it is stored, who can access it, whether it can be used to train commercial models, and what safeguards reduce over-surveillance. Preparation programs can build this competence by teaching candidates to interpret privacy policies, evaluate vendor claims, and communicate data practices clearly to students and families.

Academic integrity is a third challenge that has moved from policy language into daily instructional design. Candidates must learn how to guide students in responsible use, teach citation or disclosure of AI assistance, and design assessments that emphasize process, not just product. Because generative AI can produce fluent work quickly, programs should train candidates to build assignments that require visible thinking, revision histories, oral explanation, and reflection that reveals understanding. Integrity in the AI era is less about catching misconduct and more about teaching transparent collaboration and protecting independent thinking.

Finally, candidates must learn human–AI collaboration in a way that preserves human agency. This means verifying AI outputs, contextualizing data with lived experience, and maintaining relationships and ethical responsibilities that AI cannot replicate. In practice, human–AI collaboration is supervision: teachers decide when to use AI, when to withhold it, how to adapt its outputs, and how to explain those choices to students. Preparation programs should frame this as a professional stance rather than a technical skill, reinforcing that AI is an assistive tool whose recommendations must remain subordinate to human judgment. Teacher preparation programs are shifting from training in technology operation to preparation in technology ethics, interpretation, and supervision.

Policy, Practice, and Future Directions

Teacher preparation programs will need stronger institutional support if AI integration is to move beyond isolated experimentation. This includes faculty development, coherent program expectations, and partnerships with schools that allow candidates to encounter AI in authentic instructional settings. AI literacy should be developed across the program in ways that connect technology use to pedagogy, ethics, and professional judgment. From chalkboards to AI, teacher education has continually evolved alongside the tools that shape classroom life. What distinguishes AI is not simply its novelty, but its capacity to influence planning, assessment, interpretation, and decision-making at the cognitive level. Teacher preparation must therefore respond not by adding one more tool to the curriculum, but by preparing educators to exercise sound judgment within increasingly AI-mediated environments.

With this historical foundation in place, Chapter 4 turns from description to design by outlining the competencies new teachers need and the structural changes preparation programs can make to develop ethical, equity-centered human supervision of AI. The shift matters because AI's influence is not confined to instructional materials; it reaches into feedback, assessment, and decision-making, where errors and bias can affect opportunity. Preparing educators for AI therefore requires more than comfort with tools. It requires program-wide expectations for verification, data and privacy boundaries, bias awareness, and the professional habit of knowing when human judgment must override automation.

Chapter 4:
Redefining Teacher Preparation

Teacher preparation has always been shaped by the tools and expectations of its era. When new technologies enter schools, they rarely change only classroom materials; they also reshape what teachers must notice, decide, and be accountable for in real time. Generative AI accelerates this pressure because it can produce instructional content and recommendations quickly, often with an authority that can feel convincing even when it is incomplete or wrong. As a result, the central question for preparation programs is no longer whether candidates can operate new tools, but whether they can supervise them with judgment, ethics, and care.

The previous chapter established that AI changes teacher preparation because it increasingly participates in work that looks like professional judgment, including planning, feedback, assessment, and decision-making. This chapter moves from that diagnosis to design. It focuses on how teacher preparation programs can be rebuilt so candidates learn to use AI productively while protecting student dignity, privacy, equity, and the human relationships that make learning possible.

AI is not simply another classroom technology. Because it can generate, interpret, and recommend, it can influence instructional choices, feedback, and assessment in ways that require supervision. Preparing teachers for this reality means teaching them not only how to use AI tools, but how to question them, verify them, and decide when not to use them (Holmes et al., 2019). Teacher preparation programs therefore need redesign, not minor adjustment, embedding AI literacy, ethics, and instructional judgment across coursework and clinical practice. The next section explains why this transformation is necessary and why AI raises the stakes for professional responsibility, equity, and trust. From there, the chapter identifies the core competencies new teachers need and the program structures that help candidates practice ethical, human-guided supervision of AI.

Why Teacher Preparation Programs Must Transform in the Age of AI

Historically, teacher preparation programs often treated technology as an addition rather than a transformation. AI changes that pattern because it can generate explanations, detect misconceptions, analyze student work, draft lessons, and recommend next steps. In practice, this means AI can influence the very moves teachers make, sometimes well, sometimes poorly, and sometimes unfairly. Because AI now participates in planning, feedback, assessment, and decision-making, it sits closer to the core of professional judgment than most prior classroom technologies. Preparation programs must therefore develop teachers who can collaborate with AI critically and ethically, preserving student agency and keeping human judgment central. As Selwyn (2019) argues, AI shifts teachers' work toward interpretation, evaluation, and oversight of machine-generated insights. These responsibilities require explicit preparation rather than informal experimentation.

Teacher education must build AI literacy as a professional responsibility. Teachers need to evaluate AI outputs, recognize bias and risk, protect student data, and support student use without weakening independent thinking. This literacy includes understanding why fluent AI output can still be wrong, how prompts and context shape what a model produces, and when verification is required before ideas become instructional materials or feedback. It also includes distinguishing low-stakes uses, such as brainstorming examples, from high-stakes uses, such as grading, placement, or discipline signals, where human review must be explicit and documented. AI literacy enables teachers to supervise AI rather than simply follow it, and it is becoming as essential as earlier literacies demanded by print and digital technologies (UNESCO, 2024).

Inadequate preparation introduces multiple risks. Teachers may rely on AI-generated outputs without sufficient critical evaluation, potentially diminishing their creativity and professional autonomy over time. Algorithms developed from biased data sets can perpetuate harmful stereotypes, misread multilingual writing or dialect, or recommend interventions that do not fit a student's lived context. There is also a risk of inadvertently exposing students to data exploitation when teachers use unvetted tools or paste identifiable student work into systems with unclear retention or training practices. Finally, families may lose trust in schools where AI is used with limited transparency, especially if decisions appear automated, inconsistent, or hard to challenge. Teacher preparation programs should therefore treat AI literacy as a professional and ethical duty, not just a technical skill.

For these reasons, teacher education must shift from teaching tool use to cultivating supervised practice. Candidates need repeated opportunities to critique AI outputs, justify instructional choices, and reflect on how AI affects equity, privacy, and relationships. The next section shows that this kind of transformation is not unprecedented: earlier technologies reshaped teacher education when they changed what teachers were expected to do, how classrooms were organized, and what counted as professional competence.

How Earlier Technologies Reshaped Teacher Education

Teacher preparation has always responded to new tools, such as chalkboards, audiovisual media, computers, the internet, and mobile devices, by adding skills and updating methods courses. The pattern is consistent: programs teach teachers how to operate new tools and incorporate them into lessons. The limitation is also consistent: programs often underestimate how technologies reshape assessment, equity, and professional responsibility. AI raises those stakes because it can participate in planning, feedback, and interpretation. The key lesson from earlier technologies is that successful adoption requires changes in pedagogy, supervision, and ethics, not just training on features.

Artificial intelligence is more than a tool teachers rely on. It plays many roles: it interprets student work, shapes instructional choices, customizes learning paths, creates content, organizes feedback, supports equity, and affects educational policy. Teacher preparation therefore requires redesign, not just expansion. As the OECD (2025) notes, AI requires educators to develop new forms of professional judgment grounded in data ethics and algorithmic understanding. Teacher education must evolve from teaching how to use tools to teaching how to supervise intelligent systems. Teacher preparation programs should equip candidates with AI-specific competencies, including understanding how AI functions, its outputs, and its effects on teaching. Six core competencies support responsible educational AI use.

1. **AI literacy (how systems work).** Teacher candidates need a working understanding of what large language models do, how training data shapes outputs, and why AI can be biased, inaccurate, or hallucinate. This is not about coding; it is about knowing enough to supervise AI use responsibly (UNESCO, 2024). Strong AI literacy also includes understanding the difference between "fluent" output and trustworthy output, and knowing when verification is required. Candidates should practice evaluating AI-generated lesson drafts, explanations, and feedback for accuracy, tone, and developmental appropriateness. Over time, the goal is to form the habit of treating AI output as a draft to be reviewed, not as an authority to be followed.

2. **Ethical literacy (how to use AI responsibly).** The ethical risks of AI in education include algorithmic bias, privacy violations, surveillance, opaque decision-making, and automation that undermines human agency. Teacher candidates must learn to identify these risks and make choices that prioritize student well-being, dignity, and equity. Ethical literacy includes the ability to question vendor claims, challenge biased outputs, recognize when a tool may marginalize certain groups, and set clear boundaries for student use and disclosure. UNESCO (2024) stresses that ethical literacy must be a core element of teacher preparation, not an elective or a one-time workshop. It requires case-based learning and repeated practice in making tradeoffs, not technical training alone.

3. **Pedagogical integration (designing learning with AI).** AI should enrich, not replace, sound pedagogy. Teacher candidates must learn to use AI to generate options for differentiation, examples, and explanations while preserving inquiry, authenticity, and student thinking. This means learning to critique AI-generated lesson plans, revise them to fit student needs, and design tasks that require sensemaking rather than quick completion. Holmes et al. (2019) argue that AI is most valuable when it amplifies human strengths such as creativity, curiosity, empathy, and moral judgment. Preparation programs should therefore train candidates to decide when AI improves learning design and when it undermines it.

4. **Data and privacy literacy (protecting student information).** AI depends on data, much of it sensitive, which means teachers must understand privacy as a core professional responsibility. Candidates should learn FERPA-aligned practices, consent boundaries for minors, and how data is stored, retained, and deleted across tools. They also need to

evaluate vendor claims about data use, including whether student work may be used to train commercial models, and what audit logs or safeguards exist. Privacy is not only legal; it is relational: families are more likely to trust schools when AI use is transparent, and data practices are clearly bounded. Programs should include practical scenarios so candidates practice making privacy-protective choices under real classroom constraints.

5. **Equity and cultural competence (detecting disparate impact).** AI can widen inequities if used without critical awareness. Teacher candidates must understand how bias enters algorithms, how AI may misinterpret multilingual writing or nonstandard dialects, and how predictive analytics can reinforce stereotypes or deficit narratives. They should learn to conduct simple equity "spot checks" of tools and outputs by asking: Who benefits? Who is harmed? Who is missing? What cultural assumptions shape the model's defaults? The OECD (2025) emphasizes that equity should be a primary lens for evaluating AI in education, not an afterthought. Programs can build this competence through structured critique of AI-generated feedback and assessment recommendations across diverse student examples.

6. **Human–AI collaboration (preserving the human dimensions of learning).** Teachers must preserve the human dimensions of learning, including empathy, rapport, relational trust, cultural responsiveness, and moral decision-making, while using AI as a brainstorming partner and generator of possibilities. Collaboration skills include knowing when AI should step in, when it should step back, and how to refine AI-generated ideas without losing teacher voice or student agency. Candidates should practice "supervision" routines: questioning outputs, verifying claims, checking for bias, and

contextualizing suggestions with knowledge of learners. This competency also includes helping students use AI ethically by teaching verification, citation/disclosure expectations, and boundaries that protect independent thinking. In short, candidates learn to work with AI without surrendering the professional and relational responsibilities that define teaching.

Developing these competencies requires structural changes to teacher education; a few workshops or add-on modules are not enough. First, AI cannot live in a single "educational technology" course. Programs need to embed AI across foundations, human development, content methods in literacy, math, and science, assessment, classroom management, special education, culturally responsive teaching, and fieldwork seminars. In literacy methods, for example, candidates can practice critiquing AI writing feedback; in STEM (Science, Technology, Engineering, Mathematics) methods, they can evaluate AI-generated explanations and problem sets; and in assessment courses, they can analyze the risks of automated scoring.

Second, clinical practice is where professional identity is formed, so programs should ensure candidates learn to use AI in planning, differentiation, formative assessment, data analysis, and communication in real classrooms. Clinical supervisors can help candidates practice a crucial skill, knowing when *not* to use AI, especially during high-stakes interpretation of student behavior, when student voice and agency are at risk, or when a task requires original reasoning.

Third, AI-integrated simulation labs can provide repeated rehearsal of challenging scenarios, including feedback-giving, culturally responsive dialogue, classroom management, and parent–teacher conversations, before candidates lead these moments with real students. Simulation environments enable targeted practice, immediate coaching, and exposure to linguistic and cultural diversity. Dieker et al.

(2014) found that simulation can improve teacher readiness, reduce anxiety, and enhance performance in real classrooms.

Fourth, preparation programs should update how they assess candidates by evaluating professional judgment in the presence of AI. This includes candidates' ability to analyze AI outputs, preserve academic integrity, design AI-resistant assessments, explain AI concepts to K–12 students, differentiate responsibly, and avoid unethical shortcuts.

Finally, faculty development and program-wide cultural change are prerequisites for sustainable AI integration. Most teacher educators were trained in pre-AI eras and need time, support, and shared resources to redesign curriculum, develop cases, and align practice with ethical expectations. Without investment in faculty learning communities and coherent program expectations, AI preparation will remain uneven and dependent on isolated champions.

The Impact of AI on Teacher Identity and Professionalism

AI can surface fears about replacement and professional worth, especially for new teachers who are still forming an instructional identity. Preparation programs should name these concerns directly and pair them with a clear counter-message: the relational, ethical, and interpretive dimensions of teaching are not automatable. Selwyn (2019) notes that anxieties about automation often mask deeper questions about professional value; programs can address this by emphasizing that teacher authority comes from context, relationships, and judgment, not speed of content generation. In practice, this means making space in methods courses and seminars for candidates to discuss what "good teaching" looks like when AI can generate explanations and materials instantly. It also means helping candidates distinguish between productive AI support and technologies that subtly pressure teachers to comply with algorithmic recommendations rather than exercise professional agency.

AI also reshapes how novices think about authority. Because tools can generate polished explanations quickly, candidates may confuse fluent output with pedagogical expertise. Teacher education should therefore help candidates see authority as something earned through interpretation, responsiveness, cultural competence, and the ability to build learning environments where students can think, not as something granted by having the "best answer" at the fastest speed. Programs can reinforce this by requiring candidates to justify when they accept, revise, or reject AI-generated instructional suggestions and by evaluating the reasoning behind those choices. Over time, these routines train candidates to treat AI output as a starting point for professional judgment rather than a substitute for it.

Creativity in the AI era shifts from generating everything from scratch to curating, revising, and designing with intention. Candidates should learn to use AI as a brainstorming partner while still protecting teacher voice and instructional coherence. For example, a candidate might use AI to draft three possible inquiry prompts for a unit, then select one, revise it to reflect local community context, and build a sequence of student tasks that requires discussion, evidence, and revision rather than quick completion. This is not a loss of creativity; it is a change in where creativity shows up, in selecting ideas, shaping tasks, and crafting experiences that fit real students. When programs make revision and justification visible, they help candidates see creativity and professional identity as inseparable from judgment.

Ultimately, teacher preparation must cultivate reflective judgment in an AI-saturated world. Candidates need habits of mind that include questioning AI's assumptions, checking for inaccuracies, noticing missing perspectives, and adapting suggestions to learner diversity. Reflective judgment also includes "verification routines," such as cross-checking sources, comparing outputs to standards, and asking what evidence would change a decision. In addition, candidates need practice noticing when ethical issues are at stake, including privacy boundaries, bias signals, and moments when AI use could reduce student agency or academic integrity. In this sense, reflective judgment becomes a core feature of professionalism, not an add-on skill.

Emerging University and College Practices in AI-Integrated Teacher Preparation

Several universities have begun redesigning literacy methods courses to incorporate AI writing feedback tools. In these models, candidates compare AI-generated comments with human feedback, look for algorithmic bias, revise AI feedback for accuracy and tone, and design mini-lessons based on patterns the tool surfaces. When faculty require candidates to justify when and how AI feedback is used, AI becomes a supervised partner rather than an authority (Holmes et al., 2019). These course designs also allow candidates to practice communicating AI use to students in age-appropriate language, which prepares them for transparency expectations in real schools. Over time, repeated comparison between AI and human feedback helps candidates develop a sharper sense of what good feedback sounds like, and what it requires from a teacher.

Other programs are building simulation laboratories where candidates rehearse high-stakes scenarios with virtual students, including culturally sensitive conversations, classroom management, and difficult parent interactions. Simulation recordings support reflective seminars and coaching, helping candidates build confidence before facing similar moments in real classrooms (Dieker et al., 2014). When simulations include AI-generated student responses, candidates can practice responding to both the instructional situation and the technology's limitations, for example, when an avatar's language is stereotyped or when the simulation nudges toward a one-size-fits-all response. These labs also provide a practical setting for coaching "in-the-moment" judgment, such as de-escalation, questioning strategies, and equitable participation moves. Used well, simulations make professional learning more consistent by giving every candidate repeated practice with comparable scenarios.

Assessment and data courses are also evolving. Where programs introduce AI dashboards and predictive analytics, candidates learn to interpret predictions cautiously, question "at-risk" labels, and balance algorithmic signals with contextual knowledge of students' experiences. The goal is not to reject data, but to prevent algorithmic certainty from replacing professional judgment. Candidates can practice looking for false precision, situations where a model's confidence seems high but the underlying data is incomplete or biased. They can also learn to document decision-making so that interventions remain transparent and accountable rather than solely data-driven.

Finally, some programs are infusing AI ethics across the curriculum by using recurring cases: bias in automated grading, consent and privacy in communication apps, attention and digital well-being, and the ethics of AI-assisted planning. Treated this way, responsible AI use becomes a habit of mind that candidates practice repeatedly, not a one-time compliance topic. Case cycles also help candidates develop a shared language for discussing risk, tradeoffs, and professional responsibility, which supports more consistent practice across instructors and placements. Over time, these repeated scenarios build moral muscle memory: candidates learn to pause, ask what could go wrong, and choose safeguards before adopting a tool. This approach aligns AI preparation with the broader professional ethics traditions already present in teacher education.

Policy Implications for Teacher Preparation Programs

Across every era, from chalkboards to computers, teacher preparation has adapted by adding new tools and skills. AI requires more than another add-on because it can shape planning, feedback, assessment, and decision-making in ways that look like professional judgment. The task for preparation programs is to graduate teachers who can use AI productively while safeguarding student dignity, privacy, and equity, and who can explain and defend when AI should not be used. Chapter 4 builds on this foundation by outlining what a redesigned teacher preparation program looks like in practice. Transforming teacher education for responsible AI use requires structural and policy shifts at multiple levels: institutional, state, national, and global. AI is too consequential to remain optional or peripheral.

Accreditation and standards bodies will need to name AI competencies explicitly. Without clear expectations, programs will vary widely in readiness, and candidates will enter classrooms with uneven preparation. Standards can specify not only what candidates should know, including basic AI literacy, privacy, and bias awareness, but what they should be able to do, including verifying outputs, designing assessments that preserve student thinking, and communicating AI use transparently to families. Clear competencies also give faculty leverage to redesign courses and clinical assessments since expectations are no longer optional or dependent on individual instructor interest. Over time, aligned standards can help prevent AI preparation from becoming a "hidden curriculum" that only some candidates receive.

State guidance also matters because teacher preparation does not operate in a vacuum. When states establish adoption guardrails, including minimum privacy protections, procurement expectations, and transparency requirements, programs can prepare candidates for the real policy conditions they will encounter in schools. This alignment helps candidates learn not only how to use AI tools, but how to work responsibly within evolving expectations. State guidance can also reduce confusion for districts and universities by clarifying what tools are permitted, what data practices are acceptable, and what training is expected before classroom use. In turn, preparation programs can build those requirements into coursework and student teaching so new teachers are not learning policy "on the fly."

Universities also need infrastructure that supports responsible experimentation, including secure data practices, faculty training, district partnerships, and regular review routines for tools used in coursework and clinical practice. Without these supports, access and quality will remain uneven across candidates and programs. Infrastructure includes practical supports such as approved-tool lists, privacy reviews for any platform used with candidate-created student data, and clear guidance on what candidates may use during fieldwork. It also includes support for faculty, such as shared case libraries, model assignments, and time to redesign courses as tools change. When infrastructure is in place, innovation becomes safer and more consistent, and candidates encounter expectations that match what districts will require.

Partnerships with K–12 systems are equally important. Programs cannot prepare candidates for AI-enriched classrooms without knowing what tools, policies, and instructional expectations local districts are using. Strong partnerships enable shared professional learning, co-designed curriculum, and feedback loops that keep preparation grounded in classroom reality. They also help align candidate practice with district guardrails, for example, what data may be entered into tools, what disclosure is expected, and what integrity rules apply to student work. When universities and districts plan together, candidates experience continuity between what they learn in coursework and what they are expected to do in schools.

Global frameworks also provide a shared ethical baseline. Guidance from organizations such as UNESCO and the OECD emphasizes human-guided AI use, equity protections, transparency, and student data rights, commitments that teacher preparation programs can use to align curriculum and expectations across contexts. These frameworks also help programs avoid treating AI as a local "tool choice" problem and instead treat it as a professional ethics issue that travels with teachers across roles and settings. When candidates learn these global commitments early, they are better prepared to interpret district policies, ask informed questions about vendors, and recognize when local practice drifts from shared ethical norms. In this way, global guidance supports both consistency and moral clarity in a rapidly changing technology landscape.

Taken together, these policy implications signal a central reality: teacher preparation cannot carry AI readiness alone. District expectations, state guidance, accreditation standards, and university infrastructure shape what new teachers are allowed and prepared to do with AI in real classrooms. When those systems align, AI integration becomes less about isolated experimentation and more about coherent professional practice grounded in ethics, equity, and human judgment. With that foundation in place, the next chapter shifts from program design to daily instruction, showing how teachers can use AI as a teaching partner in lesson planning, differentiation, and feedback, while preserving the relational work that makes learning possible.

Chapter 4 focuses on capacity building, the knowledge, habits, and program structures that prepare educators to supervise AI ethically. Chapter 5 turns to enactment, what teachers do with that capacity when they plan instruction and make day-to-day choices about tasks, scaffolds, and evidence of learning. The shift from preparation to practice is where AI's promise becomes real, and where the risks of shallow automation must be actively managed through professional judgment.

Part III:
AI in Instruction and Student Learning

Chapter 5:
AI as a Teaching Partner

Lesson planning has long been recognized as one of the foundational practices of effective teaching. It requires teachers to translate curricular standards, content knowledge, instructional strategies, and contextual understanding into a coherent learning sequence. It is also time-consuming. National survey and review evidence suggests that teachers devote substantial time to planning and preparation, both during and beyond the school day (Education Week Research Center, 2022; Tamez-Robledo, 2024; TNTP, 2025). AI can reduce parts of that workload, but its significance goes beyond efficiency: it changes how quickly teachers can prototype lessons, differentiate materials, anticipate misconceptions, and respond to evidence of learning.

Many educational technologies promised to transform instruction, yet in practice they often added tools without changing the core work of teaching (Cuban, 2018). AI is different because it can generate drafts, adapt explanations, and analyze patterns in ways that intersect directly with planning, feedback, and differentiation. That is why it is best understood as a teaching partner, useful when guided by clear goals and professional judgment and risky when treated as an authority.

Unlike earlier educational tools, contemporary AI systems can generate text, analyze patterns, scaffold learning, and adapt to student needs in real-time. These capabilities make AI more than a delivery tool; they position it as a planning partner that can reshape how teachers design instruction. In that sense, AI invites educators to reconsider planning not as a procedural task, but as a central part of effective teaching.

Too often, instructional planning is treated as an administrative requirement rather than as a vital tool for shaping learning and assessment. AI can ease this burden by generating draft plans aligned to standards, differentiating materials, and suggesting resources. Effective teachers, however, treat AI-generated content as a starting point rather than a finished product. Used well, AI supports teachers as co-designers, helping them to work more efficiently while preserving the judgment and intentionality that strong instruction requires. This chapter explores how AI can make teaching more collaborative and humane while strengthening personalization for students with varied needs and abilities.

Lesson Planning as Cognitive and Emotional Labor

Lesson planning has traditionally been a time-intensive activity requiring teachers to align standards, differentiate instruction, anticipate misconceptions, and integrate appropriate assessments. Although many educators view planning as essential to effective teaching, it is also one of the most time-consuming aspects of the profession (Creagh, 2025). Teachers often spend hours preparing units, curating resources, and adapting materials. These tasks, while intellectually meaningful, can also contribute to burnout.

Traditional lesson planning requires teachers to interpret and prioritize standards, identify learning goals and success criteria, select or create activities, anticipate misconceptions, design formative assessments, differentiate for diverse learners, and integrate social-emotional and cultural considerations. This complexity makes lesson planning both intellectually demanding and emotionally taxing. Teachers often describe planning as "invisible labor," essential to outcomes yet undervalued in policy and school structures (Evans & Martin, 2024). AI offers potential relief because it can act as a planning partner by reducing repetitive tasks while preserving space for human creativity and judgment. This matters because some of the strongest influences on student achievement remain tied to the quality of instruction, teacher clarity, and responsive teaching rather than to efficiency alone (Hattie, 2009).

AI lesson generators, such as ChatGPT, Khanmigo, Eduaide.AI, and NotebookLM, can create draft lesson plans aligned to standards, recommend activities, and differentiate materials by reading level or modality. These systems allow educators to rapidly prototype instructional ideas. Effective teachers, however, treat AI output as a starting point, not a final product. AI accelerates the drafting process but cannot replicate the teacher's contextual understanding of their students, community, or learning environment. For example, AI might generate a generic lesson on climate change. A teacher might then adapt it by incorporating local environmental data, designing inquiry tasks reflecting student interests, integrating Indigenous or community perspectives, and embedding authentic assessments aligned to school goals. AI thus functions as a planning partner rather than an autonomous planner.

Evidence suggests that educators can maintain professional control when integrating AI. A survey conducted by OECD found that 68 percent of teachers using AI lesson tools reported higher efficiency, and 83 percent modified at least half of AI-generated content, indicating strong professional oversight (OECD, 2025). These findings demonstrate the enduring centrality of teacher judgment. AI contributes efficiency and generative power, but decisions about accuracy, cultural relevance, rigor, and alignment remain human responsibilities.

The Essential Role of Human Judgment in AI Integration

Unquestionably, AI can produce coherent output quickly, but it lacks lived experience, pedagogical reasoning, and ethical judgment. Teachers must evaluate AI-generated materials for accuracy and factual correctness, alignment with developmental levels, cultural responsiveness, potential bias or stereotypes, and appropriateness of tone and context. As Gay (2018) emphasizes, culturally responsive teaching requires human sensitivity, not algorithmic pattern recognition.

AI systems reflect the patterns of the data on which they were trained. This can inadvertently reproduce stereotypes or omit marginalized perspectives. Teachers play a vital role in ensuring that AI-assisted materials remain inclusive and empowering. Human review is thus an ethical necessity, not a best practice. Balancing automation with professional expertise remains central to the planning process. While it is true that AI excels at repetition, teachers excel in areas such as relationship-building, motivating students, designing meaningful learning experiences, interpreting subtle cues, making ethical decisions, and fostering belonging and connection that are beyond the capabilities of AI. An effective integration model positions AI as a tool that liberates teachers to focus on these uniquely human dimensions.

AI as a Tool for Instructional Differentiation

Designing lessons that account for differences in student needs and abilities remains one of the most challenging aspects of instructional planning. Educators have long recognized that differentiated instruction is foundational to equitable learning and improved student outcomes (Tomlinson, 2014), yet it is also one of the most demanding expectations placed on teachers. To differentiate effectively, teachers must individualize tasks, pacing, scaffolding, grouping, and levels of challenge for dozens of students simultaneously; a task that requires deep pedagogical expertise, ongoing assessment, and significant planning time. In many classrooms, the scope and complexity of differentiation exceed what is realistically possible for teachers to achieve alone daily.

AI-powered adaptive learning systems attempt to bridge this gap by managing portions of the diagnostic and responsive workload. Platforms such as CenturyTech, Knewton Alta, and DreamBox use machine learning models to analyze patterns in student performance, identify misconceptions, and recommend individualized practice pathways (Pane et al., 2015; Woolf, 2021). These tools offer tiered activities, adjust the level of challenge in real time, provide targeted remediation, and offer enrichment options for advanced learners. By rapidly analyzing student data at scale, AI enhances educational equity by making individualized support more feasible and more consistent across diverse classrooms. Importantly, however, personalization must remain grounded in professional judgment. Algorithms can suggest what students *might* need; teachers, drawing on relational knowledge, professional judgment, and assessment, determine what they *truly* need (Holmes et al., 2019).

A major feature of AI's role in instructional planning is its ability to generate personalized learning pathways and deliver real-time instructional support. AI tutors, chat companions, and digital learning assistants can provide on-demand help by answering questions, rephrasing explanations, offering strategic hints, and modeling step-by-step problem-solving approaches. These capabilities align closely with Vygotsky's (1978) theory of the zone of proximal development, which posits that learners benefit from just-in-time guidance that extends their ability to solve complex tasks. AI is a dynamic scaffold, helping students persist through challenges while maintaining appropriate cognitive demand.

Beyond supporting content mastery, AI systems increasingly foster metacognitive and self-regulatory development. Some platforms prompt students to ask clarifying questions, reflect on their strategies, identify misconceptions, set goals, and monitor their progress over time, which are skills essential to 21st-century learning and lifelong problem-solving (Azevedo, 2015). Emerging empirical evidence supports this potential. Recent research summarized by Burns (2026) suggests that generative AI tutoring can improve learning and engagement when paired with human instructional support. Their findings suggest that AI can significantly enhance cognitive engagement, especially when embedded within thoughtful human instructional design rather than used as a stand-alone intervention.

Improving Accessibility and Equity with AI

AI also helps address the diversity of learners present in today's classrooms by generating individualized pathways and adaptive learning plans tailored to each student's needs, abilities, and prior knowledge. This approach aligns with the principles of Universal Design for Learning (UDL), an educational framework that encourages teachers to design instruction that is flexible, inclusive, and responsive to learner variability (CAST, n.d.; Meyer et al., 2014). UDL minimizes obstacles by emphasizing engagement (the "why"), representation (the "what"), and action and expression (the "how"). This approach ensures that all students, irrespective of their language background, disability status, or learning profile, have equitable access to the curriculum and can demonstrate their understanding (Shultz, 2023). The goal is to cultivate "expert learners" who are resourceful and knowledgeable, strategic and goal-directed, and motivated and purposeful (CAST, n.d.). AI-powered platforms can further support UDL by offering customizable resources and real-time feedback that adapt to each learner's progress. This dynamic personalization helps students overcome challenges and fosters greater independence and confidence in their educational journey.

AI-supported accessibility tools further extend these principles by offering adaptive, multimodal supports that remove barriers and promote equitable participation. Systems such as Microsoft Immersive Reader, Google's Read Along, and advanced speech-to-text or text-to-speech programs provide robust assistance for English language learners, students with dyslexia, learners with visual or auditory impairments, and students who experience processing or attention-related challenges (Al-Azawei et al., 2016; Meyer et al., 2014). These tools can translate or simplify text, provide real-time read-aloud functionality, highlight key vocabulary, and offer customizable font sizes, contrast settings, and pacing controls. Many tools also adjust dynamically to student behavior by slowing down speech when a learner hesitates or offering structured prompts when comprehension appears to lag (Holmes et al., 2019).

These AI-driven supports matter in two ways. First, they expand access to instructional materials that may otherwise be difficult for some learners to engage with, reducing structural inequities in the classroom (Ok et al., 2022). Second, they empower students to interact with content more independently by offering individualized, just-in-time accommodations without requiring constant adult intervention. This independence aligns with the core objectives of UDL and inclusive pedagogies, which emphasize building learner agency and reducing reliance on one-size-fits-all interventions (Meyer et al., 2014). In doing so, AI helps normalize the use of assistive technologies, reducing stigma and making differentiated supports available to all students rather than only those with formal accommodations. This wider availability of accessibility tools, consistent with UDL principles, helps diverse learners participate more fully, demonstrate their strengths, and build confidence as autonomous learners.

AI as a Tool for Experiential and Dynamic Learning

AI enables dynamic learning environments where students apply knowledge in authentic contexts rather than merely absorb information. These environments can include virtual science laboratories, AI-generated economic models, and historically grounded role-play scenarios. AI can also support civic debates, social science inquiry, and adaptive social-emotional learning scenarios

(Holmes et al., 2019; Luckin et al., 2016). These tools deepen conceptual understanding by requiring students to test ideas, interpret feedback, and revise their thinking (Dumont et al., 2010).

For example, in a high school social studies course, students might use AI to "interview" historical figures whose responses draw on large collections of primary documents, enabling learners to probe motives, biases, and contextual constraints (Schmidt & Cohen, 2013). They might also participate in AI-mediated civic dilemmas, where the system adapts arguments based on student reasoning, prompting them to consider trade-offs, consequences, and alternative viewpoints (Holmes et al., 2019). Similarly, students could explore counterfactual socioeconomic scenarios, such as modeling the effects of different tax structures or migration policies to see how changes ripple through a society (Holmes et al., 2019). These experiences cultivate critical thinking, empathy, and perspective-taking by placing students inside complex systems and encouraging them to reason through uncertainty rather than memorize static facts (Luckin et al., 2016; Dumont et al., 2010).

AI as a Tool for Contextualization and Pedagogical Reflection

While AI can generate instructional materials with impressive speed, it lacks the situated, experience-based knowledge that teachers develop over years of practice. Shulman's (1987) concept of pedagogical content knowledge highlights this essential human dimension: effective teaching requires not only mastery of subject matter, but also insights into students' prior knowledge, common misconceptions, cultural backgrounds, motivations, and the specific contexts in which learning occurs.

These relational and contextual understandings are built through sustained human interactions with students, parents, and communities, and thus cannot be automated or fully replicated through computational models (Darling-Hammond et al., 2020). As scholars of culturally responsive pedagogy emphasize, localizing instruction to reflect the lived experiences and cultural identities of students increases relevance, strengthens engagement, and improves learning outcomes (Gay, 2018). AI can assist with certain technical aspects of planning, but the interpretation, adaptation, and humanization of instruction remain squarely within the teacher's expertise.

AI can support teachers' planning, especially when its suggestions are subjected to careful, systematic review. One such area is cognitive load. By scanning lesson plans, instructional sequences, or student-facing materials, AI can help teachers identify where explanations may be overly dense, where multiple new concepts are introduced simultaneously, or where extraneous details may interfere with comprehension. These insights support the application of cognitive load theory, which emphasizes managing intrinsic and extraneous load so students can devote working memory to core learning processes (Sweller et al., 2019). Rather than replacing teacher judgment, AI functions as a diagnostic aid that calls attention to areas where learners may struggle.

AI tools can also support the anticipatory elements of pedagogical reasoning. Teachers can prompt AI to generate likely misconceptions associated with a topic, drawing on research about common errors in mathematics, science, literacy, and other domains. This aligns with evidence from formative assessment research showing that anticipating misconceptions allows teachers to design targeted questions, frameworks, and feedback that more effectively surface and address student thinking (Heritage, 2018). By simulating student misunderstandings, AI provides a kind of "rehearsal space" in which teachers can refine explanations, clarify conceptual sequences, and adjust instructional moves before presenting them in the classroom.

Finally, AI can assist with evaluating alignment across objectives, instructional activities, and assessments. Teachers often struggle to ensure that these components fit together coherently, especially when planning under time constraints. Assessment scholars argue that alignment is essential for instructional coherence because it ensures that students are being taught and assessed on the exact knowledge and skills intended in learning outcomes (Popham, 2017). AI-based alignment checks do not replace deep professional knowledge of standards or student needs, but rather, they offer a structured prompt for teachers to reflect on whether their plans logically support the learning goals they intend to measure. In all these ways, AI serves not as a substitute for academic wisdom but as a reflective tool that enhances teachers' ability to plan thoughtfully, anticipate challenges, and make intentional instructional decisions grounded in context and professional judgment.

Risks and Limitations in AI-Assisted Planning

While AI offers meaningful benefits for instructional planning, its integration also introduces significant risks and limitations that require careful professional judgment. One of the most promising advantages of AI is its potential to free teacher time for the relational and reflective dimensions of teaching. Research consistently shows that increased opportunities for meaningful feedback, one-on-one conferencing, and strong teacher–student relationships correlate with higher academic achievement and improved social-emotional outcomes (Pianta et al., 2020; Shute, 2008). By letting AI handle routine tasks like creating practice problems, writing lesson introductions, or recommending materials for different learning needs, teachers can spend more time on complex educational thinking and engaging directly with students in ways that technology can replace.

AI systems, however, also produce algorithmic deviations, commonly referred to as "hallucinations," in which the model generates incorrect, misleading, or fabricated information presented in a confident, authoritative tone (Crawford, 2021). In instructional contexts, this poses a significant risk to student learning. AI-generated lessons may contain conceptual inaccuracies, incomplete structures, inappropriate sequencing, or misleading examples. These issues demand vigilant teacher verification. Educators must review, revise, and contextualize AI outputs to ensure accuracy and alignment with standards and student needs.

Another limitation concerns the nature of AI-generated content itself. While AI can produce coherent and polished lesson materials, such outputs can lack depth, inquiry orientation, or alignment with constructivist learning principles. As decades of learning science research demonstrate, effective instruction requires designing opportunities for students to build understanding through exploration, discourse, and problem solving (Bransford et al., 2000). AI often defaults to teacher-centered, procedural, or surface-level explanations unless prompted with significant specificity. Without careful adaptation by educators, AI-designed activities may inadvertently reinforce passive learning rather than promote the deeper, conceptually rich engagement that students need.

Equity concerns further complicate AI's role in educational planning. Unequal access to reliable technology, high-quality digital infrastructure, and ongoing professional learning can amplify pre-existing disparities among schools and districts. Scholars warn that AI systems often reproduce the social and racial biases embedded in the data on which they are trained, disproportionately impacting marginalized communities unless safeguards are intentionally implemented (Benjamin, 2019). International organizations echo these concerns, noting that the rapid adoption of educational AI without robust equity-focused policies, guidelines, and training risks widening opportunity gaps both within and between nations (UNESCO, 2023). Ensuring that all students benefit from AI-supported instruction requires deliberate investment in teacher preparation, digital literacy, and infrastructure.

In Chapter 5, AI functions best as a co-designer: drafting materials, generating options for differentiation, supporting accessibility, and surfacing patterns, while teachers verify accuracy, strengthen rigor, and ensure cultural and developmental fit. Used without oversight, AI can introduce errors, default to shallow activities, or reproduce inequities; used with professional judgment, it can free time for the relational work and instructional decision-making that students most need. Chapter 6 builds directly on this point by shifting from planning to learning: how teachers can use AI to create the conditions for deeper student understanding, including critical thinking, metacognition, transfer, and meaningful engagement, rather than faster completion of tasks.

Chapter 6:
AI as a Catalyst for Deeper Student Understanding

Chapter 5 argued that AI can make planning faster, but it cannot make planning wise. The same is true of learning. AI can accelerate practice, generate explanations, and deliver feedback, but deeper learning still depends on how teachers design experiences that require students to think, revise, and transfer understanding. In other words, AI can increase the speed and volume of academic activity, but it cannot guarantee that students are building durable concepts, flexible strategies, or the habits of mind that endure beyond a single assignment.

This distinction matters because generative tools can produce fluent answers that look like understanding. A student can submit a polished paragraph, a correct solution, or a confident explanation and still be unable to explain why it works, diagnose their own errors, or apply the idea when the context changes. Deeper learning shows up when students can make their reasoning visible, compare and critique approaches, revise ideas in response to feedback, and use concepts in unfamiliar situations. It is revealed through the learning process: the drafts, the discussion, the missteps, and the refinements, not merely the final product.

AI can support deeper learning when it is used to strengthen, not bypass, the core mechanisms of understanding: timely feedback, productive struggle, reflection, and application in varied contexts. It can help teachers increase the frequency of formative cycles—draft, feedback, revision—by generating practice variations, offering alternative explanations, and supplying prompts that require students to justify choices and examine assumptions. AI can also help teachers respond to learner variability by adapting scaffolds, language supports, and entry points, while still holding the cognitive demand of the task steady. Used this way, AI becomes a tool for designing better learning conditions, not a substitute for student thinking.

At the same time, AI raises real risks that can quietly flatten learning: overreliance, shallow task completion, biased or inaccurate outputs, and the temptation to outsource sensemaking to a system that cannot be held accountable. Preventing these outcomes requires active teacher oversight and transparent norms for students. Oversight includes setting clear boundaries for when AI may be used and for what purposes, requiring students to show intermediate thinking (notes, annotations, rough drafts, solution paths), checking AI outputs for accuracy and bias, and designing tasks that make learning visible through discussion, artifacts of revision, and structured reflection. This chapter examines the learning-science foundations of deeper learning, the mechanisms through which AI can support cognitive and metacognitive growth, cross-disciplinary classroom applications, and the instructional responsibilities that keep human judgment central.

Understanding Deeper Learning

Deeper learning refers to a comprehensive constellation of cognitive, interpersonal, and intrapersonal capabilities that enable students to move far beyond superficial recall of information. At its core, deeper learning is demonstrated when students can apply knowledge in complex, real-world situations; analyze, synthesize, and evaluate information from multiple sources; and transfer understanding across diverse contexts and domains. This transferability, often considered the hallmark of true understanding, reflects the learner's ability to reorganize, reinterpret, and reapply concepts flexibly rather than merely recite them. Transfer is particularly significant as it shows that knowledge gained can be used outside its initial context and adapted to new and unfamiliar situations (Perkins & Salomon, 1992).

According to the National Research Council (2012), deeper learning is underpinned by three interdependent dimensions. First, cognitive competencies involve critical thinking, problem-solving, reasoning, and robust content knowledge. These are not isolated skills but interconnected processes that require students to engage with disciplinary practices, manipulate ideas, and construct meaning. Second, interpersonal competencies, such as communication and collaboration, emphasize the social dimensions of learning. Students demonstrate deeper learning and understanding when they can articulate their thinking, negotiate meaning with peers, and engage effectively in group problem-solving. Third, intrapersonal competencies include metacognition, persistence, self-regulation, and academic mindset. These competencies help learners navigate challenges, reflect on their cognitive processes, set goals, and sustain effort, all of which are essential for mastering complex tasks.

These dimensions align closely with the well-established framework for 21st-century skills that underscores the importance of adaptability, creativity, and lifelong learning. They also align with broader conceptions of deep learning that emphasize creativity, collaboration, citizenship, and authentic problem solving beyond the classroom (Fullan & Langworthy, 2014). In contemporary global and technological contexts, the ability to continuously learn, unlearn, and relearn is increasingly vital. Deeper learning is an essential framework, equipping students to excel within a dynamic and data-driven environment.

Research in cognitive sciences shows that deep learning occurs when students actively engage, practice repeatedly, and receive timely feedback (Sawyer, 2014). Recent systematic reviews likewise suggest that large language models can support personalization, feedback, and access while also raising concerns about accuracy, bias, and overreliance that demand careful oversight (Shi et al., 2026). This distinction is important because deeper learning depends not only on exposure to content but also on active, constructive, and interactive engagement that requires students to explain, generate, and refine understanding (Chi, 2009). Bransford et al. (2000) highlight that effective learning is learner-centered, addressing prior knowledge and culture; knowledge-centered, focusing on concepts; assessment-centered, offering ongoing feedback; and community-centered, promoting collaboration and inquiry. Achieving these conditions is often challenging in traditional classrooms.

AI tools, in many respects, offer a scalable way to operationalize these principles by creating dynamic and data-informed systems that adapt to individual learners. Through predictive analytics, natural language processing, and real-time feedback mechanisms, AI can support continuous assessment, personalize instructional pathways, and help students engage more deeply with disciplinary ideas. For example, adaptive learning systems can identify when a student is developing misconceptions, while intelligent tutoring systems can prompt reflective questioning to strengthen metacognitive awareness.

Despite widespread agreement on the importance of deeper learning, achieving it at scale remains challenging. Teachers frequently face constraints related to time, large class sizes, diverse learning needs, and limited instructional resources. Conventional assessment methods typically emphasize accuracy rather than the quality of reasoning, which can limit students' ability to exhibit their conceptual understanding. Additionally, instructional approaches may inadvertently overemphasize coverage of content at the expense of exploration, inquiry, or reflective practice.

How AI Catalyzes Deeper Student Understanding

AI can deepen student understanding when it reshapes how learners engage with content, receive feedback, reflect on strategies, and apply ideas across contexts. By leveraging adaptive algorithms, natural language tools, and generative models, AI can create learning environments that are more responsive and interactive than traditional one-size-fits-all routines. The key is instructional design: AI should be positioned to prompt explanation, sensemaking, and transfer, not to supply answers that students never learn to justify.

One starting point is personalization, which underlies many AI-driven learning opportunities. Personalized and adaptive learning environments create a natural transition from conventional, one-size-fits-all instruction to dynamic systems that respond directly to learner needs. AI's ability to tailor teaching methods is one of its most important impacts on advancing learning. Adaptive learning systems modify difficulty levels, pacing, and task types based on real-time data about a student's performance. This tailoring allows learners to progress through content at a pace that matches their cognitive readiness (Walkington, 2013). Research demonstrates that personalized learning environments lead to higher student achievement, especially when instruction is aligned with individual learning profiles and goals (Pane et al., 2015; Bulger, 2016).

AI-powered adaptive systems analyze vast amounts of learner data, such as time on task, error patterns, strategy use, and conceptual misunderstandings, to create individualized learning pathways. These systems also provide just-in-time interventions that prevent misconceptions from solidifying. By offering timely support and corrective feedback, AI supports learners in building interconnected knowledge structures that foster conceptual depth rather than surface-level memorization.

Building on adaptive supports, intelligent tutoring systems (ITS) deepen learning by providing structured guidance. This transition from broad personalization to fine-grained tutoring highlights how AI can shift from adapting content to actively shaping the learner's inquiry process. Intelligent tutoring systems (ITS) emulate the strengths of one-on-one human tutoring, which has long been recognized as one of the most effective instructional approaches (Bloom, 1984). ITS platforms such as Carnegie Learning's MATHia, ALEKS, or AI-driven writing assistants offer step-by-step guidance, context clues, feedback, and opportunities for reflection.

Decades of research confirm that ITS can produce learning gains comparable to, and sometimes exceeding, those achieved through traditional classroom instruction (VanLehn, 2011). These systems dynamically assess student thinking, detect misconceptions, and adjust instructional sequences accordingly. Their responsiveness enables scaffolded inquiry, where learners are guided to explain reasoning, test hypotheses, and reflect on strategies, key components of deeper learning (Koedinger & Corbett, 2006).

While intelligent tutoring systems foster inquiry, AI also plays a key role in helping students develop the internal habits that support deeper learning. Metacognition, thinking about one's own thinking, is essential for deeper learning because it enables students to monitor, evaluate, and adjust their cognitive processes as they engage with challenging material. This transition from externally guided learning to internally regulated learning reflects a developmental shift in which students gradually assume greater responsibility for their academic growth. AI tools play a critical role in supporting this shift by embedding metacognitive prompts and reflective opportunities directly within learning environments.

As learners engage with content, AI-powered systems can prompt them to reflect on their reasoning, articulate their strategies, and assess their confidence levels. These prompts act as cognitive "checkpoints," helping students pause, evaluate their understanding, and make necessary adjustments before moving forward. For instance, systems may ask students to explain why they chose a particular solution path, compare alternative approaches, or assess how confident they feel about an answer. This reflective practice fosters awareness of learning strategies and strengthens the ability to self-correct.

AI-powered engagement detection systems further enhance metacognitive development by identifying signs of frustration, confusion, disengagement, or rapid guessing. These systems analyze indicators such as response time, error patterns, hesitation, and keyboard click behavior to infer when students are struggling or losing focus. By delivering timely nudges, such as reminders to re-read instructions, suggestions to try a different strategy, or prompts to take a brief pause, AI supports learners in regulating emotions and behaviors that influence cognitive performance over time.

Learning analytics dashboards provide another layer of support by visualizing progress, achievement patterns, and performance areas that require further attention. These dashboards help students identify strengths, pinpoint challenges, and track improvement over time. This ongoing visibility aligns with Zimmerman's (2002) framework of self-regulated learning, which emphasizes iterative cycles of forethought through planning and goal setting, performance through strategy use and self-monitoring, and self-reflection through evaluation and adaptation. When students interact with these dashboards, they deepen their metacognitive awareness and engage more intentionally with the learning process.

AI tools also facilitate the development of academic mindsets that sustain persistence through personalized feedback that highlights and affirms effort, strategy use, and progress. AI reinforces growth-oriented beliefs that encourage learners to view challenges as opportunities rather than obstacles. By combining reflective prompts, emotional awareness, and strategy support, AI helps cultivate resilient learners who are better equipped to navigate complex tasks.

AI-driven metacognitive scaffolding creates a learning environment where students develop both the skills and dispositions necessary for deeper understanding. As learners become more adept at monitoring their thinking, evaluating strategies, and adapting approaches, they build autonomy and agency: qualities that not only enhance academic performance but also support lifelong learning.

AI in Practice: Cross-Disciplinary Applications

Across the curriculum, teachers and students can benefit from the thoughtful incorporation of AI into instruction. For example, in language arts classes, AI writing assistants strengthen writing skills by offering real-time feedback on structure, clarity, evidence, and tone. Natural Language Processing (NLP) tools also provide summarized readings, vocabulary suggestions, and tailored reading-level adjustments. Research on this topic indicates that AI can improve student writing by reducing cognitive load and enabling students to focus on meaning-making (Roscoe & McNamara, 2013). Furthermore, AI reading platforms like Amira or Lexia support early literacy development through phonemic recognition and fluency coaching.

Students learning mathematics benefit from AI-driven personalized problem sets that provide step-by-step instructions and conceptual visualizations. AI can also identify specific procedural or conceptual errors made by students and intervene accordingly. For example, automated detection of error patterns helps teachers understand whether students struggle with computation, conceptual reasoning, or interpretation of word problems.

AI supports scientific inquiry through simulation tools, virtual labs, and data-analysis platforms. Students can model complex systems, such as climate change, genetics, or chemical and physical phenomena, using AI-assisted simulations to test hypotheses and predict outcomes. These experiences deepen conceptual understanding by situating learning within authentic scientific practices. In the Humanities and Social Sciences, AI tools help students analyze historical documents, identify rhetorical strategies, explore ethical dilemmas, and examine socio-cultural patterns. Generative AI can simulate dialogues with historical figures or generate multiple perspectives on complex civic issues, fostering critical thinking and empathy.

AI can support deeper learning when it is used to increase opportunities for explanation, feedback, revision, and transfer, while teachers protect the integrity of the learning process. The practical challenge is not whether AI can generate strong responses, but whether students are asked to do the thinking that makes understanding durable: making claims, testing ideas, revising work, and applying learning in unfamiliar contexts. When AI is treated as a partner in that process, it can expand access, personalize support, and make learning more responsive; when it is treated as a shortcut, it can weaken independence and deepen inequities. Chapter 7 extends this argument by examining how AI interacts with major learning frameworks and by introducing the AI-Augmented Learning Cycle (A2LC) as a model for designing AI-supported learning that remains human-centered and intellectually rigorous.

Chapter 7:
AI and the Evolving Landscape of Learning

It was argued, in the previous chapter, that AI supports deeper learning only when it increases opportunities for explanation, feedback, revision, and transfer, without replacing the thinking students must do. This chapter extends that argument by asking a practical design question: What does it look like to align AI use with well-established theories of learning? To answer, the chapter connects AI to major learning frameworks and then introduces the AI-Augmented Learning Cycle (A2LC) as a usable model for designing AI-supported learning that stays human-centered and intellectually rigorous.

In K–12 schools, AI is already influencing lesson design, student work, and instructional decision-making. For educators and school leaders, the central issue is not whether AI will be present, but whether its use strengthens learning or quietly undermines it. AI can extend capacity (drafting materials, generating practice, and surfacing patterns), yet it must remain subordinate to teacher judgment, student agency, and equitable practice (Brynjolfsson & McAfee, 2017).

A key reason AI matters for learning is its ability to expand opportunities for application, practice that requires students to use knowledge in varied contexts. Experiential learning theory helps explain why this matters: "Learning is the process whereby knowledge is created through the transformation of experience" (Kolb, 1984). When guided well, AI can generate scenarios, simulations, prompts, and feedback loops that make application and transfer more frequent and more responsive than traditional one-size-fits-all routines.

AI also expands the ways students can demonstrate understanding through creative application. Students can use generative tools to explore options, test ideas, and refine products in writing, design, art, music, and engineering. The central skill is not "letting AI create," but directing and critiquing AI output toward a meaningful goal: using it to extend thinking rather than replace it.

To connect classroom uses of AI to learning theory, it helps to start with a simple premise: students do not all think, reason, or self-regulate in the same ways at the same time. Cognitive development research explains how learners grow in their capacity to reason, transfer, and reflect, and why scaffolds must be developmentally appropriate. That foundation makes it possible to see where AI can genuinely support learning, e.g., clarifying, prompting explanation, providing feedback, and where it can short-circuit it. From there, the chapter turns to learning taxonomies and culminates in the A2LC framework as a design model.

AI's Role in Cognitive Development

Cognitive development refers to the lifelong progression of an individual's ability to think, reason, learn, and solve problems. It reflects both the accumulation of knowledge and qualitative changes in how individuals interpret and engage with the world (Bjorklund & Causey, 2018; Siegler et al., 2022). In classroom terms, this development becomes visible as learners expand and revise their schemas—assimilating new information into existing understanding while also accommodating experiences that require them to reorganize what they know (Piaget, 1952). Over time, growth in working memory, attention, language, and metacognition supports increasingly complex forms of thinking and self-regulation (Diamond, 2013). With guidance and practice, students also become more strategic problem solvers, learning to plan, compare alternatives, and evaluate outcomes using progressively more effective approaches (Kuhn, 2009). Importantly, development is not only a matter of "more knowledge." It also involves qualitative shifts in how thought is structured and organized, changing the way learners represent problems, connect ideas, and reason about evidence (Case, 1992).

Jean Piaget's theory remains a foundational reference point for how reasoning changes over time. Piaget (1952, 1970) described four broad stages that, while not perfectly uniform across individuals, help educators think about developmental readiness. In the sensorimotor stage (birth–2), young children learn primarily through action and perception, and object permanence gradually emerges as they begin to understand that objects continue to exist even when out of sight. In the preoperational stage (2–7), symbolic thinking expands rapidly—children use language, images, and play to represent ideas—yet their reasoning is often intuitive and centered on a single aspect of a situation (Piaget & Inhelder, 1969). During the concrete operational stage (7–11), learners develop more consistent logical reasoning in concrete contexts, including improved classification, conservation, and perspective-taking. In the formal operational stage (12+), students become increasingly capable of abstract and hypothetical reasoning, although the degree to which this develops varies by domain and experience and is strongly shaped by schooling and opportunities to reason (Keating, 2004).

Later research emphasizes that cognitive development is shaped by culture, language, schooling, and social interaction; learners do not progress in identical ways or at identical times (Rogoff, 2003; Vygotsky, 1978). This matters for AI integration: a tool's scaffold must fit the learner's current reasoning and literacy demands, and teachers remain responsible for deciding when support helps and when it becomes a crutch.

Because students vary in developmental readiness, educators often use learning taxonomies to design tasks at appropriate levels of complexity and to describe what "deeper" understanding looks like. In this chapter, Bloom's Taxonomy, Webb's Depth of Knowledge

(DOK), the SOLO Taxonomy, Marzano's New Taxonomy, Fink's Significant Learning taxonomy, and the Community of Inquiry (CoI) framework serve as lenses for examining how AI can support learning (cognitively, metacognitively, socially, and affectively) when guided by sound instructional design.

Bloom's Revised Taxonomy (Anderson & Krathwohl, 2001) builds on Bloom's original hierarchical model of educational objectives, which first organized learning outcomes from basic recall to more complex forms of thinking (Bloom, 1956). Its hierarchical structure (Remember, Understand, Apply, Analyze, Evaluate, Create) corresponds closely to the cognitive developments described by Piaget. In early and middle childhood, children's reasoning is predominantly concrete. They excel in tasks emphasizing remembering, understanding, and applying, especially when these tasks are rooted in tangible materials or familiar situations. These learning experiences align with the cognitive capabilities typical of Piaget's preoperational and concrete operational stages, in which students interpret information primarily through perceptual means and begin to develop foundational logic.

As students move toward adolescence and enter the formal operational stage, they become more capable of engaging in the higher-level cognitive processes represented at the top of Bloom's hierarchy. Activities requiring analysis, evaluation, and creation depend on students' emerging ability to manipulate abstract ideas, consider multiple variables simultaneously, and reason hypothetically (Piaget, 1970; Keating, 2004). Thus, Bloom's Taxonomy provides a developmental roadmap that helps teachers scaffold instruction from foundational to advanced cognitive tasks, ensuring alignment between instructional expectations and students' cognitive capacities.

While Bloom's Taxonomy categorizes cognitive processes, the SOLO Taxonomy, developed by Biggs and Collis (1982), describes levels of understanding, ranging from surface-level recall to sophisticated conceptual integration and generalization. The SOLO Taxonomy's five levels: prestructural, unistructural, multistructural, relational, and extended abstract map closely onto the qualitative transformations Piaget identified.

Early learners often function at the prestructural or unistructural levels, demonstrating limited or single-aspect understanding that corresponds to early developmental reasoning. As students mature cognitively, their ability to integrate multiple pieces of information grows, reflected in multistructural or relational responses. These align with the logical but concrete capabilities of the concrete operational stage. Most notably, the extended abstract level of the SOLO Taxonomy parallels the higher-level reasoning characteristic of the formal operational stage, wherein students generalize principles and apply them to novel contexts. Because the SOLO Taxonomy is outcome-oriented, it provides teachers with a precise way to assess the depth of students' understanding, regardless of age or grade level. SOLO is both an evaluative and instructional tool, guiding educators in supporting students' movement from fragmented to integrated conceptual understanding.

Marzano and Kendall's (2007) New Taxonomy of Educational Objectives offers a more comprehensive model by incorporating three interconnected systems: the Cognitive System, the Metacognitive System, and the Self-System. Each system includes hierarchical processes that collectively influence how students acquire, regulate, and use knowledge. The Cognitive System outlines processes ranging from retrieval and comprehension to analysis and knowledge utilization. These processes align with both Piagetian descriptions of cognitive development and the levels of Bloom's and SOLO. For example, retrieval and comprehension are more consistent with the concrete reasoning characteristic of younger learners, while knowledge utilization, including decision making and problem solving, corresponds with formal operational capabilities.

The Metacognitive System involves setting goals, monitoring progress, and selecting strategies. Research consistently demonstrates that metacognition develops gradually, emerging in rudimentary forms during childhood and becoming increasingly sophisticated in adolescence (Flavell, 1985). Marzano's taxonomy gives educators clear goals to develop these abilities, acknowledging that metacognitive regulation boosts students' ability to handle intricate cognitive challenges.

Fink's taxonomy expands the theoretical framework by incorporating human and relational aspects into the learning process. Fink (2003) states that meaningful learning combines foundational knowledge with application, personal relevance, and ongoing learning ability. Learning becomes significant when it integrates disciplinary knowledge, promotes self-awareness, fosters empathy, and cultivates the disposition for lifelong learning. AI contributes to these dimensions through adaptive learning pathways, authentic simulations, socioemotional supports, and reflective journaling tools. The shift from Marzano to Fink highlights how AI enriches not only cognition and regulation but also the affective and holistic dimensions of meaningful learning.

The Community of Inquiry (CoI) framework contextualizes learning within collaborative and discursive environments. The use of artificial intelligence enhances teaching presence through automated facilitation, strengthens cognitive presence through inquiry prompts and summarization tools, and supports social presence by mediating communication and fostering community. This transition from Fink to CoI reveals that significant, holistic learning becomes most powerful when embedded within a socially constructed environment, one that AI helps sustain and enrich.

Finally, the Self-System emphasizes motivational, emotional, and belief-oriented factors that shape learning. This component parallels sociocultural theories asserting that cognitive development is inseparable from cultural practices, emotional experiences, and interpersonal interactions (Rogoff, 2003; Vygotsky, 1978). For educators, acknowledging these influences is critical for designing environments that support not only cognitive growth but also motivational growth.

Taken together, Piaget's developmental theory and the taxonomies presented elucidate a coherent structure for understanding and guiding student learning. Piaget provides insight into the developmental readiness of learners; Bloom's and SOLO articulate the cognitive demands and depth of understanding required by instructional tasks; and Marzano's taxonomy integrates the cognitive, metacognitive, and affective dimensions that shape learning in real-world classrooms. This integrated approach helps teachers align tasks with student development, use taxonomies to move learning from basic to advanced levels, embed metacognitive supports, consider motivation and culture, and create environments that foster growth. Instruction responds to students' current abilities and encourages ongoing development.

The synthesis of these taxonomies shows that learning encompasses a multifaceted interaction among cognitive processes, task requirements, structural organization, regulatory approaches, personal relevance, and social engagement. Artificial intelligence can support learning across all these domains, functioning as a cognitive scaffold, structural integrator, metacognitive and motivational support, facilitator of meaning-making and identity formation, and collaborative mediator. Artificial intelligence functions both as a tool and as a mediating layer that links multiple dimensions of learning.

Pedagogical Implications of Cognitive Development

Cognitive development is a dynamic, lifelong process through which learners refine their ability to think, reason, solve problems, and make informed decisions. In the classroom, teachers encounter daily evidence of this developmental progression: young children discovering patterns, older students evaluating arguments, and adolescents engaging in abstract reasoning. While developmental theories like Piaget's contribute valuable perspectives on students' cognitive development, learning taxonomies such as Bloom's, Marzano's, and the SOLO frameworks equip educators with effective tools for designing instructional strategies that foster and enhance this growth. When combined, these perspectives help educators understand not only what students are capable of at different stages but also how to strategically structure learning experiences that promote higher-level thinking.

Cognitive development unfolds through both quantitative gains, such as acquiring more information and more skills, and qualitative shifts, new ways of thinking. Piaget's theory emphasizes that children's thinking evolves through four stages: sensorimotor, preoperational, concrete operational, and formal operational each characterized by increasingly sophisticated reasoning abilities (Piaget, 1952). These developmental transformations influence how students learn, how they interpret instruction, and how they approach academic challenges. Teachers often observe inconsistencies in learner behavior, such as when a child who can solve a concrete math problem struggles when the same concept is presented abstractly, or when an adolescent capable of rich discussion falters when asked to transfer knowledge to a new context.

These patterns reflect the interaction between developmental readiness and instructional demand. Learning taxonomies give teachers language and structure to diagnose these demands and guide appropriate supports. When viewed together, Piaget's developmental stages and contemporary learning taxonomies provide a comprehensive map for designing instruction that is developmentally responsive, cognitively rigorous, and emotionally supportive. This framework helps teachers intentionally structure learning that nurtures the full spectrum of cognitive growth, from basic knowledge acquisition to creative, abstract problem solving.

Contemporary pedagogy necessitates moving instructional design beyond linear progression toward the management of dynamic learning environments. Artificial intelligence enables the dynamic modification of cognitive complexity, supporting the integration of pedagogical frameworks like Bloom's Taxonomy with Depth of Knowledge (DOK). It fosters structural integration and conceptual consistency in accordance with the Structure of Observed Learning Outcomes (SOLO) model.

AI also promotes self-regulation and strategic reasoning, consistent with Marzano's framework, and supports personal relevance and emotional engagement as outlined by Fink. Furthermore, it sustains collaborative knowledge construction in line with the Community of Inquiry (CoI) framework. Artificial intelligence strengthens educators' ability to design lessons that accurately match cognitive expectations with students' developmental stages, while delivering specific scaffolding to facilitate progress toward higher-order thinking. Effective instructional design requires an equitable balance between AI-driven automation and professional judgment, ensuring ethical practice and responsiveness to diverse cultural contexts.

AI as a Partner in Deeper Learning

The story of teaching has always been a story of adaptation, from chalkboards to overhead projectors, from textbooks to laptops, and from quiet rows to active learning environments. Each shift brings new opportunities and new challenges, requiring educators to rethink how learning happens and how best to engage students in a rapidly changing world. Few developments have reshaped learning as profoundly as artificial intelligence. Teachers and administrators now work with students who have more information than ever before but often struggle to use it effectively. The digital age has placed endless resources at students' fingertips, yet mastering the skills to critically analyze, synthesize, and apply this information remains elusive for many. This gap between possessing knowledge and using it meaningfully remains one of education's enduring challenges. As teaching methods change with advancements in technology, educators must focus on guiding students beyond the memorization of facts, fostering their ability to achieve meaningful understanding and apply knowledge in real-world situations.

AI transforms learning by enabling students to engage, practice, and absorb knowledge in novel ways. It emerges not as a replacement for thinking but as a partner in it. Its ability to model reasoning, individualize practice, generate examples, and prompt reflection makes it uniquely suited to supporting applied learning. Educational research consistently emphasizes that learners demonstrate true understanding only when they can transfer and apply what they learn. The cognitive foundations that support this observation include the ability to make sense of new information (encoding), retrieve information from memory, apply learning to new contexts (transfer), and monitor understanding (metacognition). These processes help explain why application is essential. Students strengthen their comprehension when they are required to use knowledge in dynamic, varied, and authentic contexts (Bransford et al., 2000).

Artificial intelligence increases the breadth and diversity of contextual applications that were once challenging to implement in traditional classroom settings. With respect to encoding, artificial intelligence enables clarification, rephrasing, and the generation of illustrative examples to enhance overall understanding. AI also enhances retrieval by encouraging low-stakes practice and inquiry. Additionally, it promotes transfer by producing unique scenarios and variations, and it fosters metacognitive development by generating reflective questions based on students' work.

The AI-Augmented Learning Cycle

The AI-Augmented Learning Cycle (A2LC) is a multi-stage learning model showing how artificial intelligence can strengthen every dimension of the learning process, from initial exposure to ultimate transfer. Rooted in decades of cognitive science and learning theory, the A2LC shows how AI can function as a cognitive partner, instructional scaffold, and feedback mechanism that enhances the depth, frequency, and quality of student learning experiences. The model consists of seven interrelated stages: exposure, clarification, practice, application, feedback, reflection, and transfer. Although these stages can be presented sequentially for explanatory purposes, the cycle itself is recursive rather than linear. Students may move back and forth among stages as they refine understanding, revisit misconceptions, and apply learning in increasingly complex contexts.

Figure 7.1

The AI-Augmented Learning Cycle

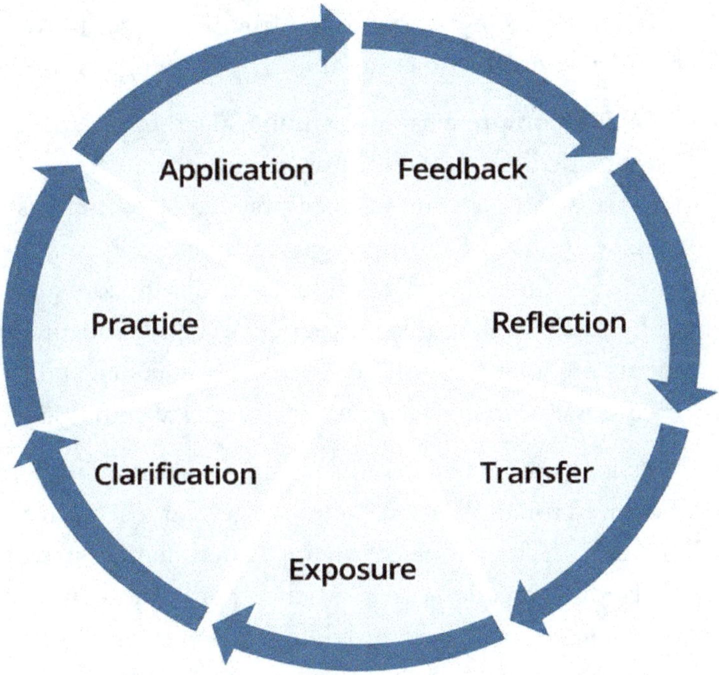

As shown in Figure 7.1, the A2LC unfolds as a dynamic seven-stage progression through which students move from initial contact with new ideas to the sophisticated transfer of learning across contexts. The cycle begins with exposure, when students first encounter unfamiliar material through reading, direct instruction, exploration, observation, or guided inquiry. At this stage, artificial intelligence is a supportive guide by clarifying unfamiliar terminology, offering concrete examples, and breaking complex concepts into more manageable components.

From exposure, students move into clarification, a stage in which they refine and verify their understanding. Here, AI becomes an interactive partner in meaning-making. It can rephrase key ideas, generate analogies, answer questions, and surface emerging misconceptions so that students can more accurately integrate new information into existing knowledge structures. Clarification helps transform initial contact with content into developing comprehension.

With initial comprehension forming, learners transition into practice, a stage dedicated to reinforcing foundational knowledge and skills. AI systems can generate tailored practice opportunities ranging from low-stakes quizzes to increasingly complex problems that allow students to build fluency and confidence at an appropriate level of challenge. Because these practice experiences can be adapted in real-time, students are able to work with greater frequency and precision than is often possible in conventional instructional settings.

As practice strengthens foundational understanding, students become prepared for application. In this stage, they use what they have learned in writing tasks, problem solving, experimentation, discussion, design work, or simulations. AI enriches application by generating scenarios, producing datasets, providing prompts, and constructing meaningful challenges that help learners recognize the relevance and utility of their developing knowledge. Application is essential for deepening and sustaining understanding; it is not just the final step in learning.

Application is followed by feedback, in which students receive information about the quality of their performance and the adjustments needed to improve. At this stage, AI tools can identify errors, model expert reasoning, suggest revisions, and provide timely, actionable responses that help learners refine their work. Feedback is especially important because it allows students to correct misconceptions before they become entrenched and to improve performance while the learning task is still cognitively active. Feedback then supports reflection, the stage in which students step back to consider what they understand well, where they continue to struggle, and which strategies have been most effective. AI can strengthen this metacognitive work by generating reflection prompts, summarizing patterns in student performance, and encouraging thoughtful self-assessment. Reflection helps learners become more aware not only of what they are learning but also of how they are learning.

The cycle culminates in transfer, where students demonstrate the deepest level of understanding by applying knowledge in unfamiliar, often interdisciplinary, contexts. Transfer requires learners to extend ideas beyond the conditions under which they were first learned and adapt them to new situations. AI plays a particularly important role here by generating varied scenarios, novel tasks, hypothetical cases, and cross-context challenges that invite learners to generalize and apply their knowledge with flexibility. Together, these seven stages create a continuous process through which students build, strengthen, and extend understanding over time.

In traditional instructional models, students often move through learning in relatively linear ways: they encounter new content, practice it, receive feedback, and eventually apply it. Yet contemporary learning, particularly in digitally rich environments, rarely unfolds in such a straight line. Students move fluidly through phases of confusion, discovery, revision, reflection, and reapplication. They revisit concepts as needed, deepen understanding through inquiry, and adapt knowledge to new contexts. The A2LC captures this recursive reality. AI enhances the cycle by providing support at every stage through examples, scaffolds, immediate feedback, modeled reasoning, simulations, and reflection prompts. What once depended exclusively on one-to-one teacher interaction can now be supplemented by responsive, personalized instructional support.

The model is deliberately cyclical rather than linear. Students may begin at different points in the learning cycle depending on prior knowledge, readiness, or task demands. A student revising an essay may enter at feedback, while a student beginning a new unit may start with exposure. As understanding evolves, learners move repeatedly through multiple stages of the A2LC, sometimes rapidly and sometimes gradually, reflecting the iterative nature of authentic learning. The cycle also reflects the principle of progressive refinement, a core idea in cognitive science suggesting that understanding becomes more accurate, flexible, and transferable through repeated practice across varied contexts. AI makes those contexts more accessible, diverse, and immediate.

Learning, as described here, involves creating meaning rather than just gathering facts. Research in cognitive psychology shows that learners strengthen and reorganize knowledge through recurring cycles of explanation, practice, problem solving, feedback, and reflection. The A2LC mirrors this natural pattern and organizes it into a repeatable framework that teachers can use intentionally. AI strengthens each stage by offering more immediate learning supports, greater personalization, more opportunities for active sensemaking, and responsive scaffolding aligned to student needs. In that sense, AI does not alter the fundamental nature of learning, but it can intensify and extend the conditions under which learning develops.

A fundamental principle of the A2LC asserts that application serves not only as an outcome of learning, but also as a key mechanism driving the learning process. Students consolidate understanding when they use knowledge to perform tasks, solve problems, or create products. Traditional constraints, including limited class time, insufficient practice opportunities, and uneven access to feedback, often restrict how much meaningful application students experience. AI reduces many of these constraints by generating tailored practice tasks, simulated environments, writing prompts, inquiry questions, data sets, revision supports, and personalized challenges. As a result, students can apply knowledge more often and more, strengthening comprehension through use.

The A2LC also places substantial emphasis on feedback and reflection, two processes consistently associated with stronger learning outcomes. AI assists by generating reflective prompts, highlighting misconceptions, modeling alternative solutions, and offering explanations aligned to student errors or partial understandings. As students move through the cycle, they become more aware of how they learn, not just what they learn. This metacognitive awareness is essential if learning is to become increasingly self-directed and transferable.

Because learning does not always proceed in predictable steps, the cycle remains intentionally flexible. A student may revisit clarification after application, cycle repeatedly through practice and feedback, or pause in reflection before attempting transfer. This adaptability makes the A2LC suitable across disciplines, grade levels, and instructional settings. It also reinforces the point that AI is most powerful not when it automates learning, but when it supports learners and teachers within a deliberately designed process of intellectual growth.

The A2LC further supports transfer through variation and simulation. Transfer, the ability to use knowledge in new and unfamiliar settings, is the clearest demonstration of robust understanding. Research suggests that transfer improves when students encounter multiple examples, practice across varied conditions, and apply ideas in contexts that differ from the original learning environment. AI is especially useful here because it can generate parallel but distinct scenarios, vary task conditions, prompt cross-disciplinary applications, and create simulations, cases, or hypothetical situations on demand. These capabilities give students more frequent and more diverse opportunities to generalize their knowledge than conventional environments often allow.

Importantly, the A2LC does not regard AI as a replacement for educators. Instead, it considers AI as a component within a comprehensive framework of instructional support. Teachers continue to develop learning experiences, interpret complex student behaviors, build classroom culture and relationships, evaluate quality and rigor, incorporate ethics and critical thinking, and nurture social and emotional growth. AI serves to provide educators with additional resources to enrich and expand student learning throughout the entire educational process.

Why A2LC Matters for Teachers

The AI-Augmented Learning Cycle (A2LC) offers educators a robust framework for understanding how artificial intelligence can enhance, rather than overshadow, the work of teaching and learning. At its core, the model clarifies an essential truth: AI functions as a scaffold that strengthens students' cognitive engagement, not a shortcut that bypasses authentic learning. By positioning AI as a support system, the A2LC model ensures that human reasoning, creativity, and judgment remain central while technology amplifies the conditions under which those capacities develop.

One of the model's most significant contributions is its emphasis on deeper learning and comprehension. The A2LC cycle's repeated practice, application, feedback, and reflection give learners more chances to engage actively with knowledge. This repeated and purposeful engagement is foundational to long-term retention. Rather than encountering an idea once and moving on, students revisit concepts in varied contexts using AI-generated examples, simulations, prompts, and feedback that enrich understanding. Over time, these layers of application help students transfer knowledge to new tasks, academic domains, and real-world challenges.

The flexibility of the A2LC model also makes it valuable across all subject areas. In language arts, AI can assist students as they draft, revise, reflect, and apply writing skills to new genres or rhetorical situations. In mathematics, AI-supported modeling allows students to explore problems, receive immediate feedback, and then extend their learning to authentic data sets. Science instruction is similarly strengthened as AI enables learners to run simulations, test hypotheses, refine explanations, and engage in iterative inquiry. Social studies teachers can use AI tools to help students analyze sources, participate in structured debates, reflect on their reasoning, and construct evidence-based arguments. In technology and engineering contexts, AI supports cycles of design, debugging, improvement, and innovation, allowing students to build increasingly sophisticated applications.

A further advantage of the A2LC model is the way it supports efficient differentiation. AI's ability to tailor instruction according to each student's needs, pace, interests, and level of preparedness at every phase allows educators to deliver personalized learning experiences while maintaining instructional rigor. Whether a student requires additional clarification, more challenging applications, or alternative modes of practice, AI can supply targeted resources that align with the learner's pathway. This individualized support frees teachers to focus on higher-level instructional decisions that facilitate discussion, provide strategic guidance, and cultivate the kinds of interpersonal connections that technology cannot replicate. In these ways, the A2LC model provides teachers with a clear, research-based rationale for integrating AI into instruction. It honors the professional role of educators, promotes deeper and more durable learning, and ensures that students across content areas benefit from responsive, differentiated, and application-rich learning experiences.

Artificial intelligence is more than an educational tool; it is a cognitive collaborator that reshapes the organization, delivery, and assessment of learning. Throughout the taxonomies discussed in this chapter, artificial intelligence is consistently identified as a multifaceted driver that promotes deeper understanding, facilitates meaningful application, and enhances integrative thinking. Rather than functioning within a single pedagogical domain, AI permeates cognitive, metacognitive, social, and affective dimensions simultaneously. This approach enhances core knowledge-building, boosts learners' abilities in analysis and problem-solving, reinforces reflective and self-regulated practices, and expands opportunities for creative and collaborative expression.

Through the lens of cognitive development, AI becomes even more significant. It aligns with decades of developmental theory by providing environments that support learners at varied stages of reasoning, from concrete experimentation to abstract, hypothetical thinking. Just as Piaget, Vygotsky, and other contemporary developmental theorists emphasize the interplay of experience, language, and social mediation, AI tools create adaptive learning spaces that meet students where they are and help propel them toward more sophisticated forms of thought. These tools can model expert reasoning, curate rich problem-solving contexts, and serve as responsive interlocutors that encourage learners to inquire, test, reflect, and revise, mirroring the very processes that drive qualitative cognitive growth.

Importantly, AI does not replace human insight or diminish the role of educators; instead, it amplifies the conditions under which deep learning becomes possible. Teachers remain the architects of learning environments, cultivating curiosity, ethical awareness, disciplinary expertise, and the relational trust that no algorithm can replicate. Artificial intelligence enables educators to tailor lessons, create practical uses, and support exploration with a level of detail and reach that was once thought impossible. In doing so, it helps bridge the long-standing gap between knowledge and real-world application, an essential hallmark of enduring understanding.

As education advances, the main goal is not just to fit artificial intelligence into established teaching methods, but to update and improve these approaches so they better reflect the growing partnership between humans and machines in learning. This shift requires educators, policymakers, and technology developers to actively rethink foundational assumptions about teaching, learning, and assessment, ensuring that educational systems remain responsive to the unique cognitive, social, and emotional needs of learners in an AI-augmented environment.

As these instructional shifts take hold, one question becomes unavoidable: how will schools recognize, document, and evaluate learning when AI can both support the learning process and generate convincing academic products? The move toward human–AI partnership does not only reshape lessons and learning cycles; it reshapes the evidence we trust. Assessment has always signaled what matters, steering student effort and teacher attention. In an AI-augmented environment, assessment must evolve from a system that primarily audits outcomes to one that can surface reasoning, growth, and authentic understanding while protecting fairness, privacy, and professional judgment. Chapter 8 turns to this next frontier by examining how AI is transforming educational assessment, what new possibilities it creates for feedback and measurement, and what safeguards are required so assessment remains a tool for learning rather than a mechanism of automation.

Part IV:
AI in Assessment

Chapter 8:
AI as a Transformative Force in Educational Assessment

Educational assessment shapes what schools notice, what students practice, and what teachers have time to respond to. For more than a century, most systems have relied on periodic measures including tests, quizzes, essays, performance tasks, and standardized exams to capture learning at a few fixed points in time. Those tools remain valuable, but they also carry familiar constraints: delayed feedback, heavy grading burdens, limited visibility into student thinking, and difficulty adapting to learner variability. In this context, assessment can feel like an interruption to learning rather than a driver of it. From a systems perspective, assessment is best understood as an evidence process-gathering, interpreting, and using information to guide learning and instructional action (Pellegrino et al., 2016).

Assessment researchers have long argued that assessment should support learning, not merely measure it (Black & Wiliam, 1998; Shepard et al., 2020). AI can help scale that vision by detecting learning patterns, interpreting student explanations in real time, generating tailored prompts, and accelerating feedback cycles. Instead of waiting for an end-of-unit test, educators can see patterns as they form and respond sooner. Large language models, machine learning analytics, automated scoring, and multimodal tools make it possible to design assessments that are more diagnostic, more personalized, and more embedded in daily instruction, if they are implemented with clear safeguards and human oversight. Automated scoring can provide rapid, consistent first-pass review of complex responses such as essays or explanations, though it also raises enduring questions about construct validity and where teacher judgment must remain decisive (Shermis, 2020). Used well, these capabilities shift assessment away from one-time products and toward evidence of process, including how students reason, revise, and build meaning over time.

For teachers and school leaders, this transformation brings both promise and responsibility. On one hand, AI can reduce the administrative burdens that have long consumed teacher time, enabling them to focus on pedagogy, relationships, and professional judgment. AI can provide students with more frequent feedback, more personalized learning pathways, and more engaging, authentic assessment experiences. On the other hand, these benefits are not automatic. AI must be implemented ethically and with attention to issues of algorithmic bias, data privacy, transparency, accessibility, and systemic inequities. Schools must ensure that AI complements, rather than replaces, teacher expertise, and that students understand both the affordances and limitations of learning with AI-powered systems.

The rise of AI forces educators to grapple with deeper philosophical questions about what is worth assessing, how knowledge should be demonstrated, and how to ensure that assessments measure meaningful learning rather than superficial patterns that AI can mimic. As generative AI becomes widely available, schools must reconsider the validity of traditional take-home assessments, the role of performance tasks, and the kinds of cognitive, social, and creative capacities that matter most in an AI-mediated world. Understanding what constitutes authentic evidence of student learning becomes more complex, and more urgent than ever.

This chapter examines how AI-driven assessment is reshaping classroom practice and school systems, with particular attention to formative assessment, feedback, evidence-centered design, and the ethical commitments that must guide implementation.

If AI is used intentionally, it can help schools shift from episodic, compliance-driven assessment toward evidence systems that are continuous, learner-centered, and growth-oriented. That shift is not automatic; it depends on design choices, professional learning, and clear guardrails that keep humans accountable for high-stakes decisions.

The Historical Context of Standardized Testing

Standardized testing rose alongside early psychometric theories and the expansion of compulsory schooling. Influenced by intelligence-testing pioneers such as Alfred Binet and by the American adoption of Intelligence Quotient (IQ) measures, educational leaders in the United States embraced the idea that cognitive ability and academic achievement could be quantified and compared (Zenderland, 1998). Standardized tests aligned with an industrial-era logic of efficiency by sorting students, standardizing instruction, and evaluating schools using common metrics. This logic promised scale and comparability, giving administrators a way to translate learning into numbers that could travel across classrooms, schools, and districts. Over time, testing helped consolidate decision-making in centralized systems by enabling ranking, placement, and accountability practices that prioritized uniform measurement over local context.

By the mid-twentieth century, standardized testing had become a central mechanism of educational accountability. Policymakers increasingly relied on test scores to make decisions about curriculum, funding, and school performance. The No Child Left Behind Act of 2001 intensified this trend, embedding standardized testing into the very structure of public education. Schools were judged on test performance, teachers were evaluated using test-based measures, and students faced significant consequences tied to their scores. This climate fostered a view of assessment centered on quantification and comparison, reinforcing the belief that learning could be captured through uniform, large-scale examinations.

Yet standardized tests have never been without critique. Scholars and educators have long argued that they offer an incomplete view of learning, often prioritizing easily measured skills over deeper understanding. They have also been criticized for narrowing curriculum, incentivizing rote instruction, and perpetuating inequities that reflect broader social stratification (Au, 2016; Koretz, 2017). Despite these concerns, the enduring presence of standardized testing speaks to its institutional entrenchment. Transforming assessment therefore requires reconsidering the assumptions behind measurement, not simply adding new tools.

The Limitations of Traditional Assessment

Standardized testing has been instrumental in structuring extensive educational systems; however, its methodological constraints limit its effectiveness in fostering genuine and fair learning opportunities. Perhaps the most fundamental critique is that standardized tests tend to emphasize recall over reasoning. Since these assessments need to be evaluated rapidly and reliably, they tend to favor clear-cut facts, procedural understanding, and set answers. As a result, students may become proficient at identifying correct answers without developing the deeper conceptual understanding necessary for long-term learning (Darling-Hammond et al., 2014). Such tests struggle to capture creativity, problem-solving, and critical thinking, skills central to success in contemporary life. Their design makes them inherently reductive, capturing only a thin slice of the rich and complex processes that constitute learning.

Another significant limitation lies in the delayed nature of feedback that standardized tests provide. Students often receive their scores weeks or months after completing the exam, long after the opportunity to revisit misconceptions or refine strategies. The separation between learning and feedback undermines the potential of assessment to support instructional decision-making. As Wiliam (2011) has argued, feedback is most effective when provided close to the moment of learning; otherwise, assessments function primarily as instruments of accountability rather than engines of growth.

Standardized tests also impose a uniform model on diverse learners. Their one-size-fits-all design presumes that all students should be assessed in the same way, at the same moment, under the same conditions. This assumption disregards the varied learning trajectories, cultural backgrounds, cognitive styles, and linguistic profiles of students. It also perpetuates inequities because students whose abilities are not captured by the standardized format may be systematically disadvantaged. The cultural and linguistic bias embedded in many standardized test items further exacerbates disparities, making these assessments imperfect measures of ability rather than reflections of systemic inequities (Ladson-Billings, 2021).

Finally, even when standardized tests are statistically reliable, they are not inherently fair. Fairness requires more than consistency; it demands that assessments reflect diverse ways of knowing, minimize cultural bias, and avoid reinforcing stereotypes. Research demonstrates that standardized tests often fall short of these expectations, producing patterns of underperformance among students from marginalized communities that reflect structural, not individual, deficiencies (Au, 2016). These limitations underscore the need for assessment systems that recognize the multidimensional, contextual, and evolving nature of learning.

The Emergence of Artificial Intelligence in Assessment

AI-enabled assessment reframes assessment as evidence collected during learning, not only after it. Systems can monitor student interactions across tasks, surface misconceptions earlier, and deliver targeted feedback while work is still in progress. In principle, this reduces the gap between instruction and assessment, though it also increases the need for transparency, bias monitoring, and careful interpretation of what the data means.

AI-enabled assessment environments operate by monitoring student behavior across learning tasks, identifying patterns that reveal strengths, misconceptions, progress, and areas for improvement. Whereas traditional assessments rely on uniform test items, AI systems adapt tasks in real-time, adjusting difficulty levels and types of prompts based on student responses. This adaptivity allows students to engage with material at a pace and level aligned with their current understanding, reducing frustration while promoting productive struggle.

Artificial intelligence enables evaluation based on relationships, allowing tools to engage with learners through dynamic conversation. When students write essays, tackle math problems, or participate in simulations, AI examines not only if their answers are correct but also how they think. It can point out flaws in reasoning, spot gaps in understanding, and recommend different strategies. Teachers benefit from ongoing feedback that helps them provide targeted and timely guidance. Meanwhile, students receive immediate responses that foster self-awareness and encourage them to embrace learning as an ongoing process.

The adoption of AI-based assessment represents a shift in educational philosophy as well as technology. Assessment shifts from classifying students to supporting continuous growth. Within this framework, learning is viewed as a sustained dialogue, with assessment integrated seamlessly into the educational process. By leveraging AI, assessment becomes an ongoing, interactive experience that adapts to each student's needs, providing timely feedback and fostering a deeper understanding of concepts. This approach encourages collaboration between students and teachers, positioning assessment as a tool for empowerment and lifelong learning rather than a mere instrument of ranking.

Continuous Feedback in AI-Supported Learning Environments

A major advantage of AI in assessment is its ability to deliver ongoing, personalized feedback to everyone. In traditional classrooms, feedback is often delayed due to the time required for teachers to review student work. AI helps mitigate this challenge by offering immediate insights while students are engaged in learning tasks. This feedback fosters a cycle of practice, reflection, and improvement, allowing students to refine understanding immediately. Feedback is most effective when it helps learners understand where they are, where they need to go, and what actions will move them forward (Hattie & Timperley, 2007).

In an AI-supported writing classroom, students might draft responses while a tool flags issues with clarity, organization, evidence, or grammar. Instead of waiting days for a teacher to review each essay, students receive suggestions they can test immediately, revising claims, strengthening evidence, or reorganizing reasoning within the same writing session. When used well, this mirrors expert practice: writing as iterative refinement rather than one-and-done submission.

Teachers remain essential to interpret AI-generated feedback against learning goals and to evaluate qualities AI cannot measure reliably, including originality, nuance, voice, and intellectual risk-taking. They also calibrate feedback to the student's developmental level, language background, and intentions, recognizing when a suggestion improves clarity and when it flattens meaning or discourages experimentation. In other words, AI can accelerate feedback, but it cannot decide what counts as strong thinking or fair evidence for a particular student. Those judgments require human knowledge of the curriculum, the learner, and the broader purposes of the assignment.

AI-supported environments extend beyond writing. In mathematics, AI-driven tutoring systems can detect exactly where a student's reasoning falters and offer tailored scaffolding. In science, simulations can adapt scenarios based on student experimentation patterns, giving learners opportunities to test hypotheses and explore consequences in real-time. For language learners, speech recognition technology allows for instant feedback on pronunciation and grammar. Across disciplines, AI's ability to analyze learning as it happens provides feedback that is more timely, actionable, and personalized than traditional assessments allow.

AI Tools and Formative Assessment

AI's greatest contributions align with research emphasizing the power of formative assessment that occurs during the learning process to inform teaching and support student progress. Numerous AI tools already in use across schools exemplify this principle. For instance, automated writing analysis platforms such as Grammarly EDU and Turnitin Revision Assistant evaluate the clarity, structure, and coherence of student writing, offering suggestions for improvement. Adaptive learning platforms adjust the difficulty and focus of tasks based on student performance, tailoring instruction to their unique learning paths. AI-driven concept mapping tools help students visualize relationships among ideas, enhancing comprehension and promoting the integration of knowledge.

These tools offer instructors data-rich insights into how students are learning and where they may require additional support. Rather than sorting or ranking students, AI tools illuminate pathways toward mastery. Their ability to provide immediate and specific feedback enhances student engagement and increases their awareness of the processes involved in developing understanding. This shift moves assessment closer to the heart of learning, emphasizing inquiry, revision, and intellectual growth.

How AI Supports Teachers

AI does not diminish the instructional role of teachers; rather, it amplifies their ability to make informed decisions. By aggregating data from multiple student interactions—exit tickets, short-response prompts, quiz items, writing drafts, and even patterns in time-on-task—AI can provide a more comprehensive view of class-level trends than most educators can assemble through manual review. Instead of relying on a handful of graded samples or intuition alone, a teacher can see which misconceptions persist, which concepts are fragile, and where students are confusing procedures with underlying ideas. For example, if students across different class periods are misapplying the same rule or misunderstanding the same concept, the teacher can revise the next day's instruction, build a focused mini-lesson, or reteach using a different representation or analogy. Likewise, if a dashboard shows that a small but consistent group is stuck at the same step in a multi-part task, the teacher can adjust scaffolds, change grouping, or offer targeted conferencing. In this sense, AI dashboards offer a level of diagnostic precision and speed that would be difficult to achieve through manual grading alone—while still requiring teachers to interpret the data through knowledge of students, curriculum goals, and classroom context.

Teachers benefit not only from the insights AI can generate but also from the cognitive and temporal relief it can provide. When routine feedback tasks are partially supported by AI—such as identifying common grammar issues, flagging unclear claims, suggesting next-step practice, or organizing student responses by misconception—teachers can redirect more of their time and energy toward higher-order instructional work. This includes mentoring students, planning rich learning experiences, facilitating discussion and inquiry, and strengthening relationships that sustain motivation and belonging. In many classrooms, AI also functions as a triage tool: it helps teachers decide where human feedback is most needed, which students require immediate conferencing, and which patterns call for whole-class reteaching rather than individual correction. Research on AI-assisted writing feedback suggests that these tools can strengthen students' writing development when they are used to support revision and improvement rather than to replace teacher evaluation. In that research, teachers reported that AI reduced their workload while enabling more timely and personalized feedback cycles (Ekizoğlu & Demir, 2025). The integration of AI into classroom practice thus contributes to a shift in the teacher's role. Rather than functioning primarily as an evaluator of final products, the teacher becomes a facilitator of growth who designs feedback loops, teaches students how to use suggestions critically, and guides them as they navigate complex learning pathways and make sense of AI-generated insights—always retaining responsibility for fairness, instructional priorities, and the humane interpretation of student work.

Preserving Human Judgment and Ensuring Fairness

While artificial intelligence enhances efficiency and adaptability, it does not possess the human attributes necessary for conducting fair and substantive assessments. Empathy, care, moral judgment, cultural understanding, and contextual awareness remain firmly within the domain of human education. "Substantive" assessment is not only a matter of scoring accurately; it is a matter of interpreting meaning and consequences—what a piece of work reveals about a learner's thinking, growth, and next steps, and how an evaluative decision might shape identity, opportunity, and motivation. Teachers therefore must interpret AI-generated feedback with sensitivity and professional discretion, recognizing when automated suggestions may misrepresent student intent, flatten voice, or disregard linguistic diversity. A tool may flag a multilingual student's phrasing as "unclear" when the idea is strong but expressed through emerging academic language, or it may mark dialect features as errors even when they are legitimate forms of expression. Similarly, teachers must attend to context that AI does not reliably "see," such as disability accommodations, trauma-informed considerations, or the instructional supports that were (or were not) available during the task.

Teachers must also evaluate student work for qualities that AI systems cannot reliably measure, such as creativity, nuance, intellectual originality, ethical reasoning, and the risk-taking that often accompanies genuine learning. In writing, for instance, an algorithm may reward formulaic structure while undervaluing narrative craft, humor, subtlety, or the careful development of an unconventional argument. In performance-based work, it may struggle to recognize inventive approaches, productive ambiguity, or culturally grounded ways of communicating meaning. Preserving human judgment means protecting these dimensions of learning from being replaced by what is easiest to quantify.

Ensuring fairness also requires that teachers monitor AI tools for unintended bias and false precision. Algorithms trained on large datasets may inadvertently reinforce cultural or linguistic norms that disadvantage certain groups, especially when models are trained primarily on dominant varieties of English, conventional essay structures, or historically inequitable disciplinary and achievement data. Practical monitoring includes comparing how feedback reads across student subgroups, conducting periodic "spot checks" using the same rubric with and without AI assistance, and investigating patterns that seem inconsistent with classroom observations. Teachers also need clear recourse procedures: when a student questions an AI-influenced score or comment, there must be a transparent path for human review and correction. The preservation of human judgment is not a rejection of AI, but a recognition of complementary roles. AI can help surface patterns and accelerate feedback, but teachers remain responsible for validity, equity, and the ethical interpretation of evidence within the lived context of learners' lives.

Ethical Considerations in AI-Driven Assessment

The effectiveness of AI-driven assessment depends on robust ethical frameworks that prioritize student rights and educational integrity. Since AI systems depend on extensive student data, maintaining transparency is crucial. Transparency means more than a general statement that "AI is in use"; students and families should be able to understand what information is collected, why it is collected, how it is used to generate scores or feedback, and what the system will and will not do well. Schools should be prepared to explain, in plain language, how human oversight functions, how students can ask questions or challenge AI-influenced judgments, and how assessment decisions remain accountable to educators. Privacy must be safeguarded through compliance with laws such as the Family Educational Rights and Privacy Act (FERPA) and the General Data Protection Regulation (GDPR), as well as through institutional policies that govern data security and limit third-party access. Ethical data governance also requires data minimization, clear retention and deletion timelines, and contracts that restrict secondary use of student work (including using student data to train commercial models) unless explicitly authorized.

Informed consent is another crucial consideration, particularly when AI tools shape evaluation, placement, or records that may follow students over time. Schools should establish clear opt-in procedures for AI-driven assessment programs, ensuring that participation is truly voluntary, that families are not pressured to consent by default, and that comparable alternatives are available for those who decline. Consent processes should specify what data will be captured (e.g., writing samples, audio, keystrokes, or interaction logs), who can access it, and how families can revoke consent or request deletion when appropriate. Ethical implementation also requires that AI be used to support learning rather than monitor or police students. The distinction between educational insight and surveillance must remain clearly defined, with explicit limits on monitoring, clear prohibitions on using assessment data for punitive purposes without human review, and safeguards that prevent "always-on" tracking from becoming normalized classroom practice.

Bias monitoring is also a necessary component of ethical AI use. AI systems should be regularly audited to ensure fairness across demographic groups, linguistic backgrounds, disability status, and cultural contexts, especially when tools score writing, analyze language, or generate risk flags. Ethical monitoring involves more than a one-time review at purchase; it requires ongoing checks for disparate impact, error rates, and patterns of feedback that systematically disadvantage specific students. In practice, this can include routine spot-checks of AI comments against teacher rubrics, subgroup comparisons of scoring distributions, reviewing false positives and false negatives, and creating a clear process for students to appeal AI-influenced results for human reevaluation. Teachers and education administrators play a vital role in identifying biases that automated systems may overlook because they are closest to students' lived language and context. Ethical AI-driven assessment therefore requires a collaborative effort involving educators, administrators, policymakers, technologists, and students, with shared expectations for transparency, documented review cycles, and clear accountability for addressing harms when they appear.

Philosophical Shifts in Education in the Age of AI

AI not only introduces new technical possibilities but also fundamentally changes how we assess education. When feedback is ongoing and learning is made transparent, schools begin to emphasize development rather than simply judging students. Students begin to understand learning as an iterative process rather than a series of high-stakes events, and mistakes become valuable data rather than indicators of failure. This shift aligns with research on growth mindset, emphasizing resilience, curiosity, and reflection. With AI-driven assessment, students can see errors and revisions as part of learning.

This environment cultivates a culture where persistence and exploration are valued, and where students develop greater self-awareness and confidence in their abilities to learn and adapt. As a result, educational experiences become more personalized, supportive, and responsive to individual needs, reinforcing the importance of lifelong learning and continuous improvement. AI also challenges the assumption that learning should follow a linear, uniform trajectory. By accommodating individualized pathways, AI supports more dynamic and adaptive models of instruction. This flexibility encourages educational systems to move away from standardized pacing guides and toward structures that honor the complexity of human development.

Perhaps the most significant shift lies in the redefinition of the teacher's role. As AI takes on analytic functions, such as processing student data, generating feedback, and identifying learning patterns, teachers are freed to become mentors, facilitators, and designers of learning experiences. In this new environment, educators have more time to engage students in meaningful conversations, encourage critical thinking, and foster creativity. Classrooms can become more humane, not less, as teachers focus on nurturing relationships, cultivating inquiry, and supporting students' emotional and intellectual growth. By prioritizing connection and individualized guidance, teachers can help students navigate complex learning pathways and develop skills that extend beyond academic achievement, including resilience, curiosity, and a lifelong love of learning.

Implications for Future Educational Systems

The integration of AI-driven assessment tools invites a reexamination of long-standing educational structures. Grading systems centered on summative evaluation may give way to more descriptive and reflective forms of assessment, enabling teachers to provide richer, more nuanced insights into student growth over time. Instead of limiting meaningful feedback to specific moments dictated by rigid assessment calendars, educators can now engage in ongoing documentation of learning, capturing student progress as it unfolds. Curriculum and instructional designs premised on uniform pacing may be reimagined to accommodate individualized learning pathways, allowing students to advance at their own speed and explore topics according to their interests and needs. This shift not only enhances personalization but also supports the development of resilience, curiosity, and a lifelong love of learning, as students experience education that responds dynamically to their unique trajectories.

For these transformations to take root, schools and school districts must invest in teacher professional development, equipping educators with the skills needed to interpret AI feedback, recognize biases, and integrate AI tools into instruction. Schools must also develop comprehensive data governance policies that establish clear ethical standards for AI use. Collaboration between educators, researchers, technologists, and policymakers will be essential to ensure that AI enhances rather than undermines educational equity. The future of assessment will depend not on the technology itself but on how schools choose to use it. AI provides the tools to rethink assessment fundamentally, but educators must decide whether those tools will reinforce existing inequities or help create more responsive, humane learning environments.

Artificial intelligence offers extraordinary opportunities to reimagine the purposes, practices, and possibilities of educational assessment. For more than a century, standardized tests have served as the backbone of measurement in schools, providing a degree of structure, comparability, and administrative efficiency that shaped the architecture of modern schooling. Yet these same systems have inadvertently narrowed the definition of learning, emphasizing what can be easily quantified rather than what is intellectually meaningful. They have often privileged recall over reasoning, rewarded uniformity over diversity, and contributed to inequities that mirror broader social disparities.

Feedback that once took days or weeks to reach students is now available in moments, enabling learners to recognize misconceptions, revise strategies, and refine understanding at the very point of need. This immediacy not only accelerates academic progress but also fosters a more reflective, metacognitive orientation toward learning, one in which students come to see growth as iterative, nonlinear, and deeply personal.

AI can help teachers identify patterns, support reflection, and provide more timely feedback, but its value depends on ethical implementation, human oversight, and a clear commitment to fairness, privacy, and meaningful learning. The future of assessment will depend less on the sophistication of AI than on the values that guide its use. Schools must decide whether these tools will reinforce narrow measurement practices or help build more equitable, reflective, and growth-oriented models of learning.

AI can make assessment more timely, more informative, and more responsive, but it also increases the consequences of interpretation. When automated scoring, predictive analytics, and real-time dashboards enter assessment systems, schools must decide which judgments can be assisted by automation and which must remain explicitly human. That boundary is where equity, validity, and student dignity are most at risk. Chapter 9 picks up this problem directly by examining how educators and leaders can balance AI-enabled efficiency with contextual expertise, ethical responsibility, and human judgment, especially when assessment data begins to influence placement, discipline, and opportunity.

Chapter 9:
Balancing Automation with Human Judgment

Teaching depends on human judgment, expressed through interpretation, empathy, and ethical responsibility exercised within real relationships and real contexts. Yet AI is increasingly used to grade work, predict student needs, recommend interventions, personalize learning pathways, and manage administrative tasks. The promise is efficiency and earlier insight; the risk is that automated outputs begin to substitute for the contextual expertise and moral reasoning that schools exist to provide. This chapter examines how educators and leaders can benefit from AI without surrendering the judgment that gives teaching its human and ethical force.

AI systems can process large volumes of data, identify patterns, and generate recommendations with remarkable speed. These capabilities promise greater efficiency and improved decision-making, offering relief from burdens that have long strained teachers' time and well-being (Williamson & Piattoeva, 2023). Yet such systems remain limited by the absence of human insight. This chapter examines how educators can use AI without surrendering the judgment that gives teaching its ethical and relational force. AI may assist educational work, but it cannot replace teachers' contextual understanding, empathy, and moral responsibility.

The distinction between human judgment and machine output becomes especially clear in the complexity of learners' lives. An AI system may flag an essay as off-topic, unaware that the writing reflects vulnerability or a breakthrough in self-expression. A pattern of missing work may be read as disengagement when the root cause is caregiving responsibilities, housing instability, or other socioeconomic pressures. Participation trackers may label a quiet student as passive even when the student demonstrates deep understanding through writing or small-group discussion. As Ferreyra and Hatcher-Mbu (2026) emphasize, AI can detect patterns and generate recommendations, but it cannot replace the human judgment needed to interpret context, weigh consequences, and make ethically responsible decisions.

Thus, the question facing contemporary educators is not whether AI has a place in schools, but how its growing presence should be understood and integrated. When deployed responsibly, AI can alleviate routine tasks, such as grading, data entry, and progress tracking, allowing teachers to invest themselves more fully in the relational, reflective, and ethical dimensions of their work. The risk, however, is that efficiency may override wisdom, or automated suggestions could replace teachers' nuanced decisions.

Balancing Efficiency with Responsibility

The appeal of efficiency in schools is neither new nor surprising. For decades, educators have sought tools that streamline administrative responsibilities, reduce workload, and provide timely insight into student performance. Artificial intelligence renews this promise with unprecedented scope. Through automated grading, predictive analytics, scheduling platforms, and early-warning dashboards, AI can manage tasks that once consumed significant teacher time. For educators already stretched thin by class size, accountability pressures, and emotional labor, these efficiencies understandably hold strong appeal (Williamson & Piattoeva, 2023).

Yet efficiency has ethical consequences. In education, even "routine" decisions carry human weight. Grades can validate effort, shape identity, and influence opportunity. Decisions about placement or discipline can alter a student's trajectory long after the moment has passed. When these judgments are mediated primarily through algorithms, they risk losing the nuance, empathy, and contextual understanding that educators bring to responsible decision-making (O'Neil, 2016).

One of the most pressing concerns lies in the potential amplification of bias. AI systems are only as fair as the data on which they are trained, and educational data, like all social data, reflects existing inequalities. Automated grading tools may inadvertently favor linguistic structures associated with dominant cultural groups. Predictive analytics can misinterpret patterns of behavior among marginalized students, especially when attendance, discipline, or participation data are entangled with socioeconomic factors. Without vigilant human oversight, AI tools may thus replicate or intensify disparities under the appearance of objectivity. As Ruha Benjamin (2019) warns, technologies marketed as neutral often "encode a politics of their own," reinforcing structures that disadvantage those already at the margins.

This reality reinforces the need for what scholars call augmented intelligence, a model in which AI supports but never supplants human judgment. Augmented intelligence positions the educator as an active interpreter rather than a passive consumer of algorithmic recommendations. In this model, teachers use AI-generated insights as starting points, not conclusions. They question anomalies, contextualize patterns, and retain the authority to make decisions rooted in ethical reasoning and human understanding. By keeping teachers in the loop, augmented intelligence preserves the relational and moral foundations of education while leveraging computational strengths (Holmes et al., 2019).

The task is to implement AI in ways that respect teachers' moral responsibilities. Efficiency must serve, rather than supersede, the responsibilities inherent in educational work. When efficiency becomes a guiding principle to the exclusion of compassion, judgment, and relational care, the purpose of schooling itself is diminished.

However, when efficiency functions as a means of reclaiming teachers' time and energy for deeper engagement with students, it becomes a powerful ally in sustaining the humanity of classrooms. Educators, administrators, and policymakers must ensure that every step toward automation is accompanied by equal attention to responsibility, transparency, and ethical accountability.

Ethical Frameworks: Keeping Humans in the Loop

As artificial intelligence becomes more deeply woven into educational practice, the need for ethical frameworks that preserve human judgment grows increasingly urgent. The concept of *human-in-the-loop* design has emerged as a guiding principle in responsible AI development, emphasizing that human oversight must remain central in any system that influences learning, assessment, or student well-being. Within the educational context, this method is more than a procedural requirement; it functions as a crucial mechanism for preserving teacher autonomy and ensuring respect for student dignity. Human-in-the-loop systems ensure that AI is a tool for professional interpretation rather than a mechanism for automated authority, reaffirming that educators remain the final arbiters of decisions that shape students' academic and personal trajectories.

Within such frameworks, AI-generated outputs are understood as provisional rather than definitive. Automated essay scoring tools, for example, can provide initial assessments that highlight common patterns, identify structural weaknesses, or flag potential areas for improvement. These tools operate with remarkable speed, processing large batches of written work in minutes, relieving teachers of the repetitive task of initial review. However, they are unable to identify the emotional nuances present in a student's voice, the importance of a personal anecdote, or the developmental progress reflected in a cautious effort to adopt a new writing style. For this reason, teacher review remains essential. By interpreting AI-generated suggestions through the lens of human understanding, teachers protect the integrity of the writing process and uphold fairness in evaluation, ensuring that the machine's judgments do not obscure the individuality of student expression (Holmes et al., 2019).

Similarly, predictive analytics systems, which have become increasingly prevalent in school districts, offer early indicators of potential academic or behavioral challenges. These models assess attendance patterns, assessment results, engagement metrics, and other data streams to forecast which students may be "at risk" of disengagement or academic decline. While such systems have the potential to direct support to students who might otherwise go unnoticed, they also carry risks. Without human interpretation, predictions can become labels, and labels can become self-fulfilling. Teachers must therefore analyze these alerts within the broader context of students' lived experiences. A flagged pattern may reveal a need for academic intervention, but it may also reflect family responsibilities, health concerns, or social pressures outside of school. When educators remain in the loop, they can transform predictive data into compassionate and informed responses, rather than allowing algorithmic conclusions to dictate actions (O'Neil, 2016).

Instructional planning provides another arena in which human-in-the-loop design ensures pedagogical integrity. AI systems capable of generating lesson plans, differentiated activities, or adaptive learning pathways can free teachers from the time-consuming task of developing every instructional component from scratch. Yet these automatically generated plans often rely on generalized templates or datasets that may not align with the cultural, linguistic, or community-specific needs of a particular classroom. Human oversight allows teachers to infuse lessons with contextual relevance, modify activities to reflect students' prior knowledge, and ensure alignment with broader curricular goals. This iterative process, within which AI suggests possibilities and teachers adapt them, creates a dynamic partnership in which efficiency and personalization coexist without diminishing professional expertise (Holmes et al., 2019).

The continual cycle of feedback and adjustment inherent in human-in-the-loop frameworks not only maintains accuracy but also fosters transparency and accountability. When teachers review and modify AI outputs, they gain insight into how algorithms process information, where their limitations lie, and how their recommendations should be interpreted. This transparency is vital in preventing overreliance on automated tools or misplaced trust in their neutrality. It encourages educators to remain critical consumers of technology, attuned to the ethical implications of delegating certain tasks to machines. Additionally, it upholds the professional standard that educational decisions should be thoughtful, situational, and based on human values rather than just on computational convenience (Williamson & Piattoeva, 2023).

Human-in-the-loop systems represent a commitment to the principle that technology should amplify, rather than diminish, the relational and moral foundations of teaching. By positioning teachers as interpreters, reviewers, and ethical decision-makers, these frameworks protect student individuality and keep schools accountable for the consequences of AI-informed decisions. AI can offer guidance, streamline tasks, and illuminate patterns, but it is teachers' wisdom and contextual understanding that determine what is fair, humane, and developmentally appropriate.

The Emotional and Ethical Role of Teachers

Artificial intelligence, regardless of its sophistication, cannot replicate the emotional intelligence or moral discernment central to the profession of teaching. While AI systems excel at identifying statistical trends or offering algorithmic suggestions, they remain incapable of understanding the human condition in the way educators do.

Compassion, intuition, and ethical deliberation are not peripheral to teaching; they are foundational. These qualities provide the emotional structure that supports academic learning and shape the moral environment in which students develop. Education goes beyond cognition; it is inherently relational and ethical, highlighting why AI cannot fully replace teachers.

Teachers engage daily in acts of empathy, often in subtle but important ways. They recognize when a student's frustration masks deeper anxiety, when a sudden drop in performance signals emotional distress, or when a moment of defiance reflects a need for connection rather than discipline. These insights cannot be programmed into an algorithm, for they draw from lived human experience, relational history, and the broader context of a student's life. Ferreyra and Hatcher-Mbu (2026) contend that AI can identify recurring errors in math assignments across a class, highlight reading skills that consistently challenge certain groups of students or alert a teacher to a student who is beginning to disengage. In schools, that interpretation depends on trust, shared experience, and sustained dialogue between teachers and students.

Ethical teaching also involves guiding students through ambiguity, a domain where humans think morally rather than mathematically. When students encounter failure, teachers help them reflect, adjust, and persevere. When conflicts arise among peers, teachers facilitate reconciliation that honors the dignity of all involved. When students struggle with identity, purpose, or belonging, educators serve as mentors who listen, validate, and support. These processes rely on compassion and moral awareness, capacities artificial intelligence does not possess. As O'Neil (2016) notes, machine-driven systems privilege efficiency and consistency, but human growth often follows non-linear, unpredictable paths that require flexibility and emotional responsiveness.

Furthermore, ethical reasoning in teaching frequently requires educators to question or reinterpret data-driven recommendations. Algorithms can suggest interventions, identify risk levels, or prioritize instructional strategies, but they cannot determine what is right for a particular student at a particular moment. Consider a situation in which predictive analytics recommend disciplinary intervention based on behavioral patterns. A teacher, aware of recent challenges in a student's home life, may choose a supportive rather than punitive approach. Similarly, an automated reading analysis might suggest that a student is stagnating, while the teacher recognizes subtle signs of increased confidence or improved comprehension that escape quantitative measurement. These divergences highlight the essential role of teacher intuition, an expertise cultivated through experience and grounded in care.

Fairness, too, is a moral construct shaped by human insight. While AI may strive for objective consistency, fairness in education requires contextual understanding. A student who arrives late each morning may not be irresponsible but may be caring for younger siblings. A student who submits assignments sporadically may be navigating economic insecurity or family instability. Teachers weigh these circumstances not as excuses but as essential considerations in shaping equitable responses. Their moral reasoning ensures that compassion tempers judgment and that discipline is administered with humanity rather than algorithmic rigidity (Benjamin, 2019).

When AI recommends a course of action, teachers must ask the fundamental ethical question: Is this right for this student, at this time, in this context? This question reflects the essence of moral responsibility in education. AI can highlight patterns and suggest possible explanations, yet it does not have the ability to define what is good or just in any given situation. Teachers, by contrast, are entrusted with making decisions that honor students' identities, protect their well-being, and support their growth. This moral stewardship is central to the educational mission and cannot be delegated to machines, no matter how advanced. By emphasizing compassion, intuition, and ethical judgment, teachers highlight their unique and essential role in education. Although AI can help improve efficiency and support well-informed choices, it is ultimately educators who interpret, provide context, and make decisions. This ensures that technology is used to promote human well-being, rather than replace the fundamental values of education.

Teachers as Designers, Coaches, and Stewards

As artificial intelligence becomes increasingly integrated into instructional environments, the professional identity of teachers undergoes a significant evolution. Rather than diminishing the teacher's role, AI shifts its emphasis, calling upon educators to enact their expertise in deeper and more intentional ways. In classrooms enriched by digital tools, teachers are no longer positioned primarily as deliverers of information or guardians of content. Instead, they become designers of learning, coaches who nurture growth through human connection, and ethical stewards who safeguard the values at the heart of education.

In the role of learning designers, teachers craft experiences that transcend the mechanical output of AI-generated suggestions. While digital systems may produce templates, categorize skills, or recommend instructional pathways, teachers weave these elements into meaningful sequences that reflect the culture, needs, and aspirations of their students. The design of learning is not a technical exercise but an act of interpretation and imagination. Educators must discern which AI recommendations align with curricular goals, which overlook the diverse strengths of the classroom, and which fail to capture the lived realities of the children they teach. This process makes teachers architects of learning environments where human creativity and machine insight coexist without compromising the humanity of the classroom (Holmes et al., 2019).

As coaches, teachers guide students through the emotional, cognitive, and developmental challenges that accompany growth. AI may provide feedback on a student's progress or highlight areas for targeted practice, but only a teacher can interpret that information in a way that honors the student's individual journey. When reviewing AI-generated feedback, teachers recognize not only what students have achieved but what they have attempted, risked, or overcome. Coaching involves strengthening confidence when data appears discouraging, offering grace when mistakes are misread as deficiencies, and encouraging perseverance during periods of uncertainty. Human coaching relies on empathy, insight, and relational trust, qualities that AI cannot emulate, regardless of its computational sophistication (Holmes et al., 2019; Ferreyra & Hatcher-Mbu, 2026).

Student data must be used ethically. Stewardship requires vigilance, discernment, and a commitment to justice. Teachers monitor how AI tools collect and interpret information to ensure that students are not reduced to data points or predictions. They scrutinize algorithmic outputs for signs of bias or inequity, mindful that decisions labeled as neutral may carry embedded assumptions reflective of broader social inequalities (Benjamin, 2019). Stewardship also involves advocating for students' privacy, ensuring that sensitive information is handled with care and that technological convenience does not erode students' rights to security and dignity. In their stewardship, teachers become the moral compass of an automated environment, ensuring that technological tools serve educational purposes rather than dictate them.

Teachers are increasingly taking on roles as designers, coaches, and stewards, making them active participants in shaping how AI is used in schools rather than simply adopting it. Although AI can help with instructional logistics, it is teachers who bring learning to life with meaning, justice, and humanity. As automation grows in education, educators guide the ethical use of technology to make sure it enhances, rather than replaces, the essential human qualities that make learning possible. The future of AI in education relies not on the machines themselves, but on the wisdom and integrity of teachers.

Reclaiming Time for Humanity

Time has long been the most strained and least renewable resource in education. Teachers frequently describe their work as governed by competing demands: instructional planning, assessment, documentation, communication with families, administrative compliance, and the unplanned emotional labor that emerges in any given day. These obligations accumulate into a pace that leaves little room for the reflective, relational, and creative dimensions of teaching that educators value most. Despite their best efforts, many teachers feel that the tasks essential for nurturing students, including listening deeply, offering guidance, and noticing subtle changes in behavior, are often crowded out by the pressing weight of routine responsibilities. Artificial intelligence offers not a replacement for human work but a potential reprieve from the mechanical burdens that consume educators' time.

When utilized appropriately, artificial intelligence can help educators reclaim valuable time by handling tasks that prioritize efficiency over interpersonal engagement. Automated grading systems, for example, can conduct initial evaluations to identify repeated patterns or common mistakes, thereby decreasing the amount of time teachers need to devote to assessing extensive collections of assignments. Data dashboards can synthesize attendance records, participation metrics, and assessment results into coherent summaries that would take educators significantly longer to compile on their own. AI-driven scheduling tools can assist in coordinating interventions, organizing learning groups, or aligning instructional resources without requiring teachers to navigate complex logistical arrays. In each of these scenarios, AI performs tasks that rely on speed, consistency, and pattern recognition, areas where computational systems excel (Holmes et al., 2019).

The purpose of automation in these contexts is not to distance teachers from their students but to clear space for more intentional engagement. With fewer hours spent on administrative tasks, teachers can devote themselves more fully to the human elements of their profession. They gain time to observe the quiet student who may be struggling beneath the surface, to check in with the learner who seems distracted or withdrawn, or to provide individualized feedback that encourages a student's confidence. They can listen to stories that illuminate a child's circumstances, mentor students through personal challenges, and build relationships that foster trust and belonging. Such experiences have a significant impact on students' emotional well-being and academic self-concept, effects that cannot be reproduced by algorithms (Holmes et al., 2019).

This reclamation of time also allows teachers to engage in deeper pedagogical reflection. Without the pressure of constant paperwork and routine tasks, educators can take a closer look at student assignments, work together with peers, and improve their lesson plans to meet the needs of a wide range of learners. Such reflection strengthens the quality of instruction and enhances teachers' professional agency. In an era when automation frequently raises concerns about dehumanization, the thoughtful implementation of artificial intelligence is an opportunity for re-humanization by enhancing educators' ability to be present, attentive, and responsive.

Furthermore, the time recovered through automation enhances the school's moral fabric. When teachers have time to notice students who might otherwise go overlooked, they enact the ethical commitments that define the profession. They can identify early signs of emotional distress, provide additional encouragement to students facing adversity, and intervene with compassion in moments of conflict. These human interventions are essential to creating equitable and caring learning environments, especially for students who rely on the relational constancy that teachers provide. By shifting routine tasks to AI systems, educators make deliberate space for the ethical and emotional work that machines cannot perform but is indispensable to student well-being (Benjamin, 2019).

AI becomes not a competitor but an enabler, an ally that supports the most human aspects of teaching. Reclaiming time represents both an advancement in logistical efficiency and a principled affirmation of the essential role of human presence within education. When technology is aligned with this purpose, it affirms the central principle that meaningful learning occurs through relationships, and that teachers, given the time to engage deeply, remain at the heart of those relationships.

The Limits of Algorithmic Objectivity

Although artificial intelligence is frequently presented as a neutral or impartial decision-maker, the idea of algorithmic objectivity deserves far greater scrutiny. In educational contexts, where fairness, dignity, and moral judgment are foundational, the belief that AI can serve as a purely objective evaluator risks misunderstanding both the nature of technology and the complexity of human learning. What appears objective in algorithmic outputs is in fact shaped by the choices, assumptions, and histories embedded within data and design. Therefore, teachers should place AI-generated insights into the wider context of relationships and ethics that they are uniquely qualified to understand.

AI's challenges are especially evident when algorithms encounter cultural or linguistic diversity. Systems trained primarily on dominant linguistic patterns may misinterpret dialects, multilingual writing, Indigenous narrative forms, or culturally embedded communication styles as deficiencies rather than legitimate expressions of identity. A student writing in African American Vernacular English or engaging in code-switching may be marked down by automated essay scoring, not because their ideas lack depth, but because their linguistic heritage falls outside the algorithm's frame of reference. Noble's (2018) work demonstrates that algorithmic systems frequently reproduce cultural hierarchies, privileging dominant ways of speaking and writing while marginalizing others. Thus, the veneer of objectivity often conceals a deeper entanglement with the social inequities embedded within the data itself.

A related limitation emerges in the distinction between statistical significance and human significance. Algorithms excel at identifying correlations unlikely to occur by chance, yet statistical relationships do not automatically translate into pedagogical or moral meaning. A predictive model may identify that declining homework completion strongly correlates with lower academic performance, but this statistical insight cannot reveal whether the student is struggling with instability at home, providing care for siblings, or coping with emotional distress. As Crawford (2021) emphasizes, data can reveal patterns at scale but often obscure the human stories behind them. Teachers use their contextual and relational knowledge to interpret student behaviors respectfully. Human significance resides in these narratives that are complex, evolving, and inherently unquantifiable.

Another challenge is the algorithmic commitment to strict consistency. While consistency is often celebrated as a virtue of AI systems, the unyielding application of rules can become ethically inappropriate in classrooms where flexibility, compassion, and developmental understanding are essential. Machine systems cannot distinguish between a student who submits late work due to disengagement and one who does so because of personal hardship. They cannot perceive when an unconventional response reflects creativity rather than error, or when a deviation signals emerging insight rather than misunderstanding. Teachers routinely navigate these distinctions, shaping their judgments not solely by rules but by empathy and contextual awareness. Pasquale (2020) argues that human professionals serve not as rigid rule-enforcers but as moral agents capable of discerning when rules should be tempered to serve justice. An algorithm, designed for procedural coherence, cannot replicate this ethical discernment.

These misinterpretations extend into the very modeling choices behind AI systems. Designers determine which variables matter, which outcomes to measure, and how to define success within a model. These decisions, though often presented as technical, reflect value judgments that can privilege measurable outputs over broader educational aims. Selwyn (2019) notes that educational technology frequently encodes narrow visions of learning, limiting opportunities for creativity, critical thinking, or emotional growth. In doing so, AI systems risk reinforcing reductive conceptions of student achievement.

Even when AI systems generate plausible recommendations, they cannot judge whether those recommendations are just, humane, or developmentally appropriate for a particular student. What appears statistically optimal may conflict with the ethical obligations of teaching. For this reason, professional judgment must remain with educators, who alone can interpret data considering human context, moral responsibility, and student dignity.

The limits of algorithmic objectivity reveal that AI's role in education must be understood as supportive rather than authoritative. Algorithms can illuminate patterns, streamline tasks, and offer preliminary insights, but they cannot grasp the cultural, contextual, or moral dimensions of human learning. Educators have the knowledge and ethical judgment to interpret data while respecting every student's dignity. Recognizing these limits is not a critique of AI's potential but a reaffirmation of the irreplaceable humanity at the heart of teaching.

Understanding the limits of algorithmic objectivity brings into sharper focus the indispensable role of the human educator. If algorithms can reveal patterns but not meaning, identify correlations but not causes, and apply rules but never discern when compassion should override consistency, then the work of interpreting, contextualizing, and ethically navigating these insights must fall to teachers. The limitations of artificial intelligence are not solely technical in nature but also encompass relational and ethical dimensions. These gaps can be addressed most effectively through the contextual insight that educators develop in their routine engagement with students. It is here, in the lived context of classroom life, that teachers exercise a depth of judgment far beyond the reach of computational systems. To fully understand the importance of this human capability, it is essential to analyze the characteristics of contextual expertise and how it empowers educators to convert data into informed judgment.

Contextual Expertise in Teaching

One of the defining characteristics of teaching, one that sharply distinguishes it from artificial intelligence, is the possession and exercise of contextual expertise. Contextual expertise means the thorough understanding teachers gain by consistently working with their students, the surrounding community, and the cultural and emotional settings where education takes place. This form of expertise does not arise from abstract rules or statistical patterns but from immersion in the lived realities of learners. It is built through years of observing individual students, interpreting subtle cues, navigating interpersonal dynamics, and understanding the social and developmental rhythms that shape educational experiences.

Contextual expertise allows teachers to perceive layers of meaning that AI systems, limited to the data they are given, cannot access.

Fundamentally, contextual expertise acknowledges that students represent more than mere data points; they are multifaceted individuals influenced by their personal histories, identities, and ambitions. A teacher knows when a student's silence is a sign of contemplation rather than confusion, when a sharp remark reflects frustration rather than disrespect, and when an inconsistent pattern of performance signals emotional strain rather than lack of ability. This interpretive ability is developed through relational knowledge acquired via dialogue, mutual experiences, the gradual establishment of trust, and empathy fostered by a comprehensive understanding of another individual over time. While AI can register patterns in behavior, it cannot understand their meaning; it can detect anomalies, but it cannot discern the narrative or reason behind them.

Contextual expertise also enables teachers to interpret learning within broader cultural, linguistic, and social contexts. A teacher familiar with a community's traditions, linguistic heritage, or collective experiences can recognize forms of expression that might appear unusual or incorrect to an AI algorithm. For instance, a multilingual student drawing upon several linguistic resources in a single piece of writing may be expressing identity, creativity, or cognitive flexibility, yet an algorithm trained on monolingual norms may misinterpret this as an error. Educators recognize that cultural factors influence communication and acknowledge that learning involves both cognitive processes and is intricately linked to social identity. Such insight allows educators to affirm student voice in ways AI systems are fundamentally unequipped to accomplish.

The temporal dimension of teaching further amplifies the importance of contextual expertise. Teachers accumulate a rich understanding of each student over time, noting shifts in mood, progressions in confidence, and responses to challenge. They understand that a child who struggles early in the year may flourish later, or that a sudden regression in performance may signal a need for emotional support rather than academic remediation. AI systems, by contrast, interpret data largely in the present tense. They lack an understanding of the developmental arcs and nonlinear trajectories that characterize human growth. Their insights are retrospective or predictive but never relationally grounded.

Contextual expertise empowers teachers to engage in moral reasoning, which is the ability to weigh decisions not solely by policy or precedent, but by what is right for a particular student in a particular moment. When an algorithm flags a student as "at risk," a teacher might interpret the same data differently, considering personal knowledge of the child's resilience, family circumstances, or recent struggles. When artificial intelligence recommends a disciplinary action based on observed behavioral patterns, educators may opt for a supportive intervention, recognizing that empathy and constructive guidance can yield more positive outcomes than punitive measures. This ethical discretion arises from context, and no model, regardless of sophistication, can or should replicate the moral imagination that emerges from human relationships.

Contextual expertise extends to instructional decisions as well. Teachers aren't just presenters of information; they create learning environments tailored to the specific needs and diversity of their students. They adjust pacing based on observed engagement, revisit concepts when students appear uncertain, and seize unexpected moments of curiosity to deepen understanding. Teachers recognize when students need challenge and when they need reassurance. AI-generated recommendations may provide valuable insights, but they cannot account for the small disruptions, interpersonal dynamics, or spontaneous opportunities that shape the flow of classroom life.

In essence, contextual expertise is the ability to see students in full human complexity. It integrates emotional attunement, cultural understanding, developmental knowledge, ethical reasoning, and professional judgment that cannot be encoded into algorithms. While AI can enhance a teacher's access to information, it cannot replicate the interpretive wisdom that arises from presence, care, and authentic human connection. It is this expertise, which is uniquely human, relational, and deeply contextual, that ensures that educational decisions are grounded not only in data but in dignity.

Acknowledging the value of contextual expertise highlights teachers' vital role in AI-supported education and prompts schools to consider how best to foster such professional judgment. If AI is to function as a collaborative tool rather than a controlling force, educational environments must be intentionally structured to preserve, and indeed strengthen, the human capacities that algorithms cannot replicate. Contextual expertise flourishes in cultures that value ethical reflection, transparency, and relational understanding, and it withers in systems that prioritize efficiency at the expense of humanity. Accordingly, the primary challenge for modern education is not merely the effective integration of AI, but rather the development of school cultures that enable educators to utilize AI responsibly and critically, while upholding the dignity of all learners. Establishing ethical guidelines is essential when developing systems that depend on expert judgment.

Framework for Ethical and Effective AI Integration

As schools navigate the integration of artificial intelligence into daily practice, the establishment of ethical guardrails becomes essential to ensure that technology enhances rather than undermines the values of education. These guardrails do not function as constraints on innovation but as structures that preserve humanity, fairness, and professional agency in environments increasingly influenced by algorithmic decision-making. Without such frameworks, there is a risk that AI systems, which are designed for efficiency and scale, may inadvertently erode the relational and ethical commitments that define effective teaching.

Ethical guardrails, therefore, act as both boundaries and protections, guiding how AI is deployed and ensuring that the ultimate responsibility for educational decisions remains with people, not machines. By providing clear guidelines, these guardrails help educators and administrators maintain the integrity of human-centered learning, even as technology becomes more pervasive in classrooms. The ethical commitments that should govern AI use in schools can be organized into five core guardrails, each designed to preserve human judgment, fairness, and professional responsibility. These principles collectively foster an environment where technology supports, rather than replaces, the relational and ethical foundations of education, ensuring that schools remain spaces for thoughtful, empathetic, and equitable learning.

Figure 9.1

Ethical Guardrails for AI

As shown in Figure 9.1, ethical AI use in schools depends on five interdependent guardrails: human oversight, explainability, accountability, equity, and professional autonomy. Together, these principles ensure that artificial intelligence remains a support for educational decision-making rather than a substitute for human judgment. They also establish the practical and moral conditions necessary for responsible implementation, making clear that efficiency, prediction, and scale must remain subordinate to the relational and ethical purposes of schooling.

A foundational pillar of these guardrails is the principle of sustained human oversight. Critical academic and behavioral decisions carry moral weight, shaping students' opportunities, identities, and futures. For this reason, no automated recommendation, whether related to grading, discipline, or advancement, should stand without human review. Teachers must be positioned not as passive recipients of algorithmic outputs but as active interpreters who contextualize insights with knowledge of students' experiences, circumstances, and intentions. As Ferreyra and Hatcher-Mbu (2026) argue, algorithms may identify patterns and generate recommendations, but only humans can interpret those outputs considering context, consequences, and ethical responsibility. Human oversight ensures that the moral considerations central to teaching remain intact.

Transparency, or explainability, forms another essential element of ethical AI use in schools. For AI to be meaningful and trustworthy, teachers and students must understand how systems arrive at their conclusions. When decisions are made by opaque algorithms, they can feel arbitrary or inscrutable, weakening confidence in both the technology and the institution deploying it. Explainability does not require every user to master technical details, but it does require that schools communicate the logic, assumptions, and limitations inherent in their AI tools. Such openness cultivates an environment in which data-informed decisions are understood rather than unquestioningly accepted, encouraging critical engagement from all stakeholders (Holmes et al., 2019).

Equally important is the preservation of accountability. Decision-making authority must remain with educators and educational institutions, not with the algorithms themselves. When AI is used to inform or support decisions, the teacher must still bear responsibility for interpreting the results, and the school must be accountable for the outcomes. This commitment prevents the displacement of blame onto systems that cannot bear moral responsibility. Accountability reminds educators that AI is a tool that is powerful, efficient, and insightful, but still subordinate to human judgment and institutional ethics (Williamson & Piattoeva, 2023).

To uphold equity, AI systems must be regularly audited for bias and accessibility. Every dataset used to train AI models reflects the patterns and inequities of the society from which it was drawn. Without careful and continuous examination, AI systems can perpetuate or even exacerbate disparities. Conducting regular audits helps prevent systemic biases in data from causing students from marginalized backgrounds to be unfairly labeled, ignored, or misunderstood (Benjamin, 2019). Equity-oriented guardrails reinforce the belief that technology must advance fairness rather than compromise it.

Additionally, professional autonomy must remain an explicit expectation. Teachers must be empowered to question or override AI recommendations without fear of penalty or dismissal. This autonomy safeguards educators' professional expertise and reinforces their responsibility to make decisions grounded in ethical reflection, contextual understanding, and human sensitivity. When teachers are encouraged to serve as partners, not subordinates, to AI systems, the technology becomes a collaborator rather than a controlling force.

These ethical principles create the conditions necessary for trust, an indispensable component of sustainable AI use in schools. Students must trust that AI systems will not judge them unfairly or reduce them to data points devoid of context. Teachers must trust that AI will support rather than replace their professional judgment. Parents must trust that their children's personal information is handled with care and integrity. Transparency is the foundation of this trust, allowing all members of the school community to understand the nature and purpose of the AI systems that shape their educational experiences.

Schools, therefore, have an obligation to communicate openly about the tools they employ, the data collected, the safeguards in place, and the ways in which AI-generated insights inform decisions. When stakeholders share in this understanding, they become collective stewards of ethical practice. They ensure accountability for both themselves and the system, treating AI as a collaborative partner guided by shared values. In these settings, AI supports rather than replaces human involvement in education.

Building a Responsible AI Innovation Culture in Schools

For artificial intelligence to enhance rather than erode the human heart of education, schools must cultivate cultures where innovation is guided by ethical reflection, collaborative inquiry, and a shared commitment to student dignity. Technology alone does not determine the character of an educational environment; rather, it is the values, norms, and relationships within a school community that shape how tools are adopted and interpreted. A human-centered approach to AI begins with school cultures that understand innovation not as the pursuit of efficiency for its own sake but as an ongoing effort to advance equity, deepen learning, and strengthen the relationships that sustain students' growth.

At the foundation of such a culture is a shared recognition that AI is not neutral. When schools adopt AI systems without critical conversation, whether for grading, monitoring attendance, recommending interventions, or shaping instructional pathways, they risk allowing automated decisions to influence student trajectories that bypass human understanding. Responsible innovation requires collective awareness of AI's limitations and potentials. Teachers, administrators, counselors, and support staff must have opportunities to explore how AI systems function, what data they draw upon, and how their outputs should be interpreted within the school's broader mission. This shared understanding creates the conditions for informed use, where algorithms support but do not overshadow the professional wisdom of educators.

Collaboration is essential for sustaining this awareness. When teachers work together to examine how AI tools influence instruction or assessment, they contribute to a culture of reflective practice. Discussions about ambiguous cases, instances when AI recommendations raise questions or seem to conflict with human judgment, become opportunities for professional learning rather than sources of confusion or compliance. These conversations help educators articulate not only what AI reveals but what it overlooks, and they clarify the ethical considerations that must guide decision-making. In such cultures, teachers feel empowered to speak openly about concerns, share experiences, and collaboratively refine their practices with an eye toward fairness, transparency, and compassion.

Leadership plays a critical role in shaping the tone and direction of responsible innovation. Administrators who model thoughtful engagement with AI, who ask probing questions about bias, privacy, and equity, signal to their communities that technological adoption is a matter of moral deliberation rather than mere operational improvement. When leaders center students' humanity in decisions about technology, they reinforce a culture where innovation is aligned with educational values rather than driven by market trends or superficial notions of modernization. Effective leaders create structures for teacher voice, ensure that policies reflect ethical commitments, and protect educators' autonomy to override AI outputs when contextual expertise suggests a different path.

A school culture based on informed human judgment prioritizes the deliberate development of trust as a foundational element. Teachers must trust that adopting AI will not diminish their agency or reduce their roles to enforcers of algorithmic recommendations. Students must trust that the data collected about them will be used responsibly, interpreted carefully, and never weaponized against their potential. Families must trust that schools will protect their children's privacy and communicate openly about the tools used to support learning. Trust grows when schools are transparent about their technological practices, when they explain clearly how AI is used, and when they demonstrate a willingness to adjust or abandon tools that undermine equity or relational connection.

Professional learning is also vital for sustaining responsible innovation. Teachers cannot be expected to navigate complex systems alone. School districts and schools must provide ongoing opportunities for educators to understand AI's capabilities, its limitations, and its ethical implications. Such learning experiences should go beyond technical training; they should invite teachers to consider how AI aligns with pedagogical goals, how it shapes their relationships with students, and how it might influence classroom culture. When professional development affirms teacher judgment and reinforces the irreplaceability of human presence, it strengthens the foundation upon which responsible innovation is built.

Moving from principles to practice requires more than enthusiasm for innovation; it requires a disciplined implementation process that protects students while giving educators room to learn. The roadmap below translates this chapter's commitments to transparency, accountability, equity, privacy, governance, and human judgment into a practical sequence of actions. It is designed to help schools and districts avoid fragmented adoption, strengthen public trust, and ensure that AI use remains consistent with the moral purposes of education.

Even as AI transforms education, teaching remains fundamentally a human activity. AI is valuable not because it can take over teaching, but because it can help teachers focus on what truly counts: building strong connections, encouraging development, and supporting students as they navigate both learning and life. Throughout this chapter, we have explored how the strengths of AI, with its inherent speed, efficiency, and analytic power, can enhance educational practice when aligned with the relational, ethical, and contextual expertise that only teachers possess. We have also examined the limitations of algorithmic systems, recognizing that data-driven insights, while valuable, can never fully capture the depth of human experience or the moral dimensions that permeate educational decision-making.

These conclusions also clarify what comes next. If AI can draft, recommend, and respond at scale, then the central work for schools is not simply adopting new tools but redefining who does what work in the learning process—and why. The arrival of AI makes roles and responsibilities newly visible: when students can consult a machine instantly, what does it mean to practice independence, authorship, and intellectual honesty? When teachers can delegate some routine tasks, what new expectations emerge for coaching, design, and ethical stewardship? Chapter 10 takes up this shift directly by examining how teacher and student roles must be redefined in AI-augmented classrooms so that technology strengthens professional judgment and student agency rather than substituting for thinking, relationship, and purpose.

Part V:
The Future of Teaching and Learning with AI

Chapter 10:
Redefining Teacher and Student Roles

The emergence of artificial intelligence in education marks a pivotal moment in the history of teaching and learning, a period in which long-standing assumptions about the roles of teachers and students are being reassessed, reimagined, and renewed. The previous chapter established the ethical, relational, and contextual foundations necessary for responsible AI adoption. Building upon those principles, this chapter turns toward the future, examining how AI transforms the very identities and responsibilities of those at the heart of education. Rather than viewing AI as a disruptive force that threatens the human dimension of learning, this chapter approaches it as a catalyst for rethinking how teachers and students engage in intellectual, moral, and creative work.

For more than a century, conventional schooling has positioned teachers primarily as transmitters of knowledge and students as recipients. The prevailing educational model, shaped by industrial-era frameworks and uniform curricula, has often restricted educators' ability to promote inquiry. Artificial intelligence is changing how knowledge is accessed, interpreted, and used. With information more abundant, accessible, and adaptable than ever before, the role of the teacher as the sole purveyor of content diminishes, giving rise to new roles centered on interpretation, design, mentorship, and ethical discernment. Teachers are increasingly called to orchestrate learning experiences, guide students through intellectual ambiguity, and cultivate the habits of mind necessary for thriving in complex, data-rich environments.

Simultaneously, AI reshapes the expectations placed upon students. In a world where machines can retrieve information, provide step-by-step guidance, and generate solutions in seconds, the value of learning shifts from recall to reasoning, from compliance to curiosity, and from passivity to agency. Students are no longer positioned merely as learners who consume information but as active participants who must learn to question, critique, and collaborate with intelligent systems. They must develop new ethical, digital, cognitive, and interpersonal literacies that enable them to discern when to trust AI, when to challenge its outputs, and how to use technological tools to extend their own thinking.

Yet these shifts are not automatic, nor are they purely technical. They depend on the creation of educational environments where human expertise and AI capabilities complement each other, where relationships and ethics remain at the center of practice, and where teachers and students are prepared to inhabit roles grounded in agency rather than dependence. Without such intentionality, AI risks reinforcing old patterns that reproduce inequities, narrow instructional practices, or diminish the relational fabric that sustains learning rather than transforming them. A redefinition of roles, therefore, must be guided by the principles established earlier: transparency, context, human oversight, ethical stewardship, and an unwavering commitment to dignity.

This chapter explores what these new roles might look like in practice. It examines the evolving identity of the teacher as a designer of learning environments, a facilitator of inquiry, and a moral guide in the age of intelligent systems. It considers how students can become more self-directed, reflective, and empowered as learners who partner with AI rather than rely upon it unquestioningly. It envisions classrooms not as spaces of technological replacement but as environments where human wisdom and machine intelligence interact in ways that elevate both.

The goal is to present a balanced perspective on artificial intelligence by neither overstating its promise nor exaggerating its potential disruptions. The intent is to develop an educational framework grounded in the belief that the essence of education is inherently human. In these ways, AI does not reduce the teacher's role; it amplifies it. Intelligent tools are intended to strengthen, rather than reduce, the integral roles of teachers and learners. The chapter advances that vision.

The Evolving Role of Teachers in AI-Enhanced Classrooms

As artificial intelligence becomes increasingly integrated into educational systems, the role of the teacher is undergoing a remarkable and necessary transformation. While early narratives surrounding AI in schools often emphasized automation and efficiency, a more mature understanding recognizes that the future of education depends not on replacing teachers, but on redefining and elevating their work. Artificial intelligence introduces advanced capabilities such as adaptive feedback, data visualization, and content generation; however, its integration shows the irreplaceable contributions of human educators, including ethical discernment, relational insight, cultural interpretation, and pedagogical innovation. Instead of diminishing the profession, AI can expand the intellectual and moral space within which teachers operate.

One of the most significant shifts in the role of teachers is the move from being primary transmitters of information to becoming designers of learning experiences. With AI able to deliver content, generate practice tasks, and provide immediate feedback, teachers are freed to focus on curating environments that promote inquiry, creativity, and deep thinking. In this regard, the teacher's work becomes more architectural than procedural: less about covering material and more about constructing meaningful pathways for students to explore. This reflects wider developments in the information environment, wherein knowledge is readily accessible, yet the need for effective guidance in interpreting it persists (Selwyn,

2019). In this regard, it is critical that teachers design learning that incorporates AI tools that never supersede human-led understanding.

Teachers increasingly serve as facilitators of intellectual growth rather than managers of academic compliance. AI can help track progress and personalize learning, but only teachers can orchestrate dialogue, guide interpretation, support critical reasoning, and respond compassionately to moments of uncertainty. Ferreyra and Hatcher-Mbu (2026) contend that while AI can generate patterns and recommendations, only humans can determine how those outputs should be understood in relation to context, consequences, and ethical responsibility. In an AI-rich environment, facilitation involves helping students navigate information critically, teaching them to interrogate AI outputs, and modeling the reflective habits of mind needed for responsible decision-making.

The teacher's role as an ethical guide becomes even more central as AI enters the classroom. While algorithms can assist with making decisions, they are unable to engage in ethical reasoning. Teachers, however, must continually weigh questions of fairness, autonomy, and equity as they interpret automated recommendations. This responsibility requires educators to understand the epistemic limits of algorithmic systems: how bias can be embedded in models (Noble, 2018), how data can misrepresent human complexity (Crawford, 2021), and how predictive analytics can shape expectations that risk becoming self-fulfilling (Rosenthal & Jacobson, 1968). In this role, teachers serve both as educational guides and as advocates for fairness in environments shaped by technology.

Teachers also assume the role of cultural interpreters. AI systems often struggle with linguistic nuance, cultural expression, and community-specific knowledge (Benjamin, 2019). When algorithms misinterpret or oversimplify student work, teachers intervene to restore meaning, affirm identity, and ensure that technology does not inadvertently marginalize diverse voices. Their cultural proficiency allows them to contextualize AI-generated outputs in relation to their students' distinct backgrounds, recognizing that learning involves cognitive processes as well as social factors and personal identity.

The evolving role of teachers requires a new form of professional confidence, one grounded not in technical mastery of AI systems but in the conviction that human expertise remains essential. Pasquale (2020) emphasizes that the future must "defend human expertise" rather than diminish it. In this future, teachers do not cede authority to machines; instead, they engage critically with AI tools, question their limitations, and integrate them in ways that reflect professional values. Professional autonomy thus becomes a safeguard against technological determinism, ensuring that AI augments rather than dictates pedagogical practice.

In these ways, AI does not reduce the teacher's role, it amplifies it, extending the teacher's impact within the classroom. The future of teaching will demand greater creativity in designing learning experiences, deeper ethical reasoning as teachers navigate complex technological landscapes, heightened relational sensitivity in supporting diverse student needs, and an expanded contextual understanding to interpret and guide the use of intelligent tools. As machines increasingly take on tasks that are routine or mechanical, such as automating feedback, streamlining data analysis, or generating instructional materials, teachers are positioned to reclaim and emphasize the deeper, human work that defines the heart of education: fostering inquiry, building trust, and nurturing critical and compassionate thinkers. Their role evolves not just in response to AI, but in a way that highlights their unique expertise, making them no less important in an AI-rich environment, but more essential than ever as guides, mentors, and architects of meaningful learning.

Student Agency, Collaboration, and Critical Thinking in AI-Enhanced Classrooms

As artificial intelligence becomes more embedded in educational practice, the expectations placed upon students expand and challenge traditional notions of learner passivity. In earlier models of schooling, students were often positioned as recipients of information, evaluated primarily by their ability to recall or replicate content. AI disrupts this paradigm, not by diminishing the role of the learner, but by inviting a reimagining of student identity as active, discerning, and ethically aware. When learning environments incorporate AI, students must engage not only with content but also with the technologies that mediate it, becoming agents in their own learning and collaborators in the ongoing interplay between human judgment and machine insight.

The first dimension of this evolving role is increased student agency. In AI-supported environments, learners have greater autonomy to direct their pace, explore topics of interest, and access personalized pathways aligned with their needs. Adaptive platforms can offer targeted practice or immediate feedback, allowing students to identify gaps and pursue growth with precision. Agency reaches further than simply making choices or controlling speed; it also encompasses building skills to evaluate feedback, establish purposeful goals, and actively manage one's own learning path. The agency also requires discerning when AI assistance is beneficial and when the deeper cognitive work of grappling with complexity is necessary. Selwyn (2019) argues that students should actively guide their learning with AI instead of simply accepting algorithmic recommendations.

In addition to agency, students must increasingly act as collaborators with AI systems and with the humans around them. Collaboration with AI involves learning how to use intelligent tools as partners in exploration: utilizing generative systems to test ideas, leveraging analytic dashboards to reflect on progress, and engaging in iterative cycles of creation and revision. Yet meaningful collaboration also requires skepticism and reflection. Students must learn to question AI's outputs, understand the limitations of its models, and recognize that technological recommendations do not carry the moral authority that human judgment provides. This stance transforms learners from consumers of automated guidance into co-interpreters of information, a role that parallels the interpretive responsibilities now expected of teachers (Crawford, 2021).

Crucially, the future of student learning in AI-rich environments demands a heightened emphasis on critical thinking. With machines capable of generating coherent answers, designing exemplars, and solving multi-step problems, the value of learning shifts toward the ability to assess, evaluate, and challenge. Critical thinking requires students to understand how AI systems are trained, what biases they may carry, and how their outputs can both illuminate and distort reality. Noble (2018) and Benjamin (2019) warn that algorithmic systems often reinforce societal inequities; students must therefore learn to recognize when AI perpetuates stereotypes, marginalizes cultural voices, or frames human behavior through reductive metrics.

Students should develop ethical literacy, which is the ability to consider the consequences of using AI when making decisions. They must consider questions such as: When is it appropriate to rely on machine-generated suggestions? How should data about oneself or others be treated? What responsibilities accompany access to powerful digital tools? These questions position students not only as learners but as future citizens navigating a society where AI shapes social, economic, and political life. As Pasquale (2020) suggests, sustaining human expertise depends on cultivating generations who can use technology with discernment and moral imagination.

Furthermore, as AI takes on more task-oriented functions, the interpersonal dimensions of student learning grow even more significant. Collaboration with peers, respectful argumentation, empathy across differences, and collective problem-solving cannot be automated. Teachers create the conditions for such interaction, but students must learn to inhabit these roles actively, developing the relational and communicative competencies that will remain essential in work and life. These unique competencies ensure that learners do not become isolated in personalized algorithmic pathways but remain engaged members of learning communities grounded in shared inquiry and mutual support.

In summary, the integration of AI into schooling demands that students develop far more than technical proficiency. They must become self-directed agents, reflective collaborators, and critical interpreters of digital systems. Students' learning should integrate cognitive sophistication, ethical discernment, cultural competence, and interpersonal abilities. Rather than reducing the significance of students, AI transforms their role by emphasizing skills that are inherently human and vital for succeeding alongside intelligent systems in contemporary environments.

Human-Centered Classroom Design for an AI-Driven World

As artificial intelligence becomes more deeply woven into educational environments, the classroom itself is undergoing transformation. This transformation is not limited to technical aspects; it also influences the social, relational, and intellectual structures within educational institutions. The classroom of the future is neither a fully automated space nor a nostalgic return to pre-digital simplicity. Rather, it forms a dynamic ecosystem where human and machine intelligence complement each other, each contributing according to their strengths. The goal is not to create a technologically saturated environment, but a human-centered one in which AI functions as a supportive layer and tool, enriching the learning experience while preserving the values and relationships at the heart of education.

At the center of this redefined system is the principle that technology must serve human purposes, not dictate them. AI tools can help streamline routines, tailor practice, and provide analytic insight, but they do not determine what constitutes meaningful learning. The teacher remains the orchestrator of experience, responsible for shaping the classroom's emotional tone, intellectual direction, and ethical climate. As Cuban (2018) notes, technological change in schools is always mediated by human judgment and institutional culture; the presence of AI does not negate this truth but reinforces it. Teachers integrate AI selectively, choosing when it enhances inquiry, when it deepens reflection, and when human interaction is indispensable.

This environment also reimagines the distribution of authority in learning. Traditionally, classrooms have been organized around teacher-directed instruction, with information flowing from the educator to the student. In an AI-supported classroom, knowledge flows in multiple directions: from teacher to student, from student to student, from AI systems to learners, and crucially from students back into the design of learning through their choices, interests, and insights. The teacher becomes less of a gatekeeper and more of a conductor, guiding a learning community in which diverse sources of knowledge and expression coexist. This shift acknowledges that AI expands access to information, but human relationships and purposeful structure are required for that information to become meaningful.

The new classroom setup changes how learning is paced and structured. AI-driven adaptive platforms allow students to move through content at individualized speeds, receiving support or challenge as needed. Yet this does not fragment the classroom into isolated learning trajectories. Instead, it opens space for teachers to create moments of collective inquiry that prompt discussions, debates, and collaborative projects that counterbalance personalized pathways with shared intellectual experiences. Personalization does not supplant community; instead, it allows for greater allocation of time toward more meaningful and genuine engagement. As Holmes, Bialik, and Fadel (2019) argue, the future of learning lies in balancing individual pathways with collaborative meaning-making.

Importantly, the classroom must remain an ethical space in which the use of AI aligns with principles of fairness, transparency, and respect for student dignity. Students must understand how AI tools operate and be encouraged to question their outputs. This critical engagement ensures that learners do not defer blindly to algorithmic authority but develop the digital literacy necessary to navigate a world where AI increasingly informs decision-making. As Noble (2018) and Crawford (2021) caution, algorithmic systems can perpetuate inequities unless their limitations are acknowledged and actively mitigated. A classroom grounded in ethical awareness teaches students not only how to use AI, but how to challenge it.

A human-centered classroom also requires careful consideration of students' emotional well-being. While AI tools can monitor engagement metrics or mood indicators, it is educators who possess the ability to discern the nuanced emotional context underlying student behaviors, such as frustration, loneliness, confidence, or joy. The classroom remains a place where relationships matter deeply, where trust is built through presence and responsiveness, and where the human need for belonging is recognized as central to learning. AI may support this work by providing timely insights, but it cannot replace the attuned understanding that emerges from sustained human connection.

Finally, the redefined classroom is characterized by flexibility and continual adaptation. As new technologies emerge, teachers and students must learn together how to navigate their possibilities and limitations. This collaborative evolution shifts the classroom from a fixed structure to a living system that is shaped by experimentation, reflection, and shared responsibility. Pasquale (2020) reminds us that defending human expertise in an AI era requires cultivating spaces where human creativity and judgment remain at the center of innovation. The classroom of the future embodies this principle, blending technological capability with human wisdom to create a learning environment that is not only more efficient but more humane.

In this evolving educational landscape, artificial intelligence enhances rather than diminishes the role of educators. Freed from repetitive administrative tasks and routine grading, teachers can allocate more time to impactful activities such as mentoring, guiding collaborative projects, and fostering critical thinking and emotional growth. AI tools help differentiate instruction and offer real-time feedback, enabling increased student engagement, greater autonomy, and the development of self-directed learning habits. As a result, classrooms become spaces where richer, more interconnected learning experiences unfold, blending personalized pathways with opportunities for meaningful community interaction. By adhering to ethical principles and focusing on human-centered design, classrooms utilizing artificial intelligence can foster effective collaboration between human and machine intelligence. This approach supports equitable access, comprehensive learner development, and maintains the essential role of human relationships and ethical responsibility within educational environments.

Preserving Teacher–Student Relationships in an Age of Intelligent Systems

As artificial intelligence becomes increasingly present in classrooms, the teacher–student relationship remains the essential anchor of educational life. No matter how sophisticated AI becomes, the relationship between a teacher and a student continues to shape the emotional climate, intellectual engagement, and moral direction of the learning environment. Far from diminishing its importance, AI's expansion elevates the centrality of this relationship, revealing what cannot be automated and underscoring the irreplaceable role of human connection in student development.

The essence of the teacher–student relationship lies in the trust that emerges from respect, presence, empathy, and the consistent recognition of a student's inherent worth. AI tools may identify patterns or generate personalized recommendations, but they cannot build trust. They cannot offer encouragement when a student feels discouraged, nor can they sustain belief in a learner's potential when the available data suggest otherwise. Trust depends on reciprocity, vulnerability, and care, qualities that arise only within human relationships. As Ferreyra and Hatcher-Mbu (2026) argue, AI systems can support decision-making, but human judgment remains essential because only people can interpret context, weigh consequences, and respond ethically to the needs of others. In education, that ethical attentiveness is inseparable from the relational work of seeing students as whole people rather than as patterns in data.

The teacher–student relationship provides the emotional grounding that allows students to take intellectual risks. Learning is an endeavor filled with uncertainty, frustration, and the fear of failure. When students feel supported by a caring adult, they are more willing to challenge themselves, explore unfamiliar ideas, and persist through struggle. AI systems may tutor or guide, but they cannot provide the emotional reassurance that comes from a human being who listens, responds with compassion, and celebrates growth. Research over decades has demonstrated the impact of teacher expectations on student outcomes (Rosenthal & Jacobson, 1968). These expectations, and the relational context in which they are expressed, shape the trajectories students come to believe are possible. A machine cannot convey belief; it can only convey probability.

This relationship also is a model for ethical engagement with technology itself. When teachers demonstrate how to question AI outputs, weigh them against contextual knowledge, and navigate ambiguous information, they implicitly teach students how to act responsibly in a world mediated by intelligent systems. This modeling is particularly important given the persuasive aura of algorithmic authority. Students must learn not only how to use AI tools but how to resist the illusion that these tools offer definitive answers. The relational dynamic becomes both instructional and ethical: teachers help students cultivate discernment, humility, and moral agency in their interactions with technology (Noble, 2018; Crawford, 2021).

Cultural understanding further deepens the relational bond between teachers and students. AI systems often misinterpret behaviors, linguistic expressions, or forms of knowledge that do not align with dominant cultural patterns (Benjamin, 2019). Teachers, however, can recognize meaning in diversity, affirm identities, and situate learning within the cultural and community contexts students bring to the classroom. This form of recognition is an essential element in fostering students' sense of belonging and preserving their dignity. For marginalized students in particular, a relationship grounded in respect and cultural awareness can counteract algorithmic misreadings, ensuring that the classroom remains a place of affirmation rather than erasure.

In an AI-influenced environment, the relational aspect of teaching requires greater intentionality. Educators must determine when to prioritize authentic human engagement over automated feedback because meaningful relationships support student motivation and well-being. They must discern when to intervene personally rather than allow systems to guide a student's progress. Equally important is creating relational moments that foster connection, belonging, and trust. These decisions require ongoing reflection and judgment. They reflect a professional ethic that places the student's humanity at the center of instructional choices, ensuring that technology serves to enhance, not diminish, the rich tapestry of relationships that define meaningful learning.

The future of teaching and learning will be defined less by AI than by the quality of teacher–student relationships. As technology becomes more present in education, these relationships will determine whether classrooms inspire students or merely automate them. A compassionate teacher can turn schools into welcoming communities where everyone feels noticed and appreciated. When educators lead with empathy, cultural sensitivity, and strong ethics, AI serves to enhance education: it enables personalized instruction and gives teachers more time for meaningful interactions.

At the heart of education is the teacher–student bond, which reminds us that education aims for personal growth, belonging, and the development of every learner, not just higher test scores. Through supportive relationships, students build academic abilities as well as confidence, resilience, and self-identity, equipping them for success both in and out of school. In the end, technology should strengthen the human aspects of teaching, guaranteeing that every student receives the care and encouragement they need along their educational path.

In the era of artificial intelligence, the changing roles of teachers and students are less about replacement than about redefining the human purpose of education. As this chapter has shown, the future of teaching and learning depends not on how quickly schools adopt new tools, but on how they integrate them into environments shaped by human judgment, moral responsibility, and relational trust. AI may accelerate access to information, personalize instruction, and illuminate patterns once hidden from view, but it cannot answer the deeper questions at the heart of education: What is worth learning? What does it mean to grow? How do we honor the dignity of every student?

These questions remain the domain of human educators, whose roles deepen in the presence of intelligent machines. Teachers become designers of rich learning experiences, facilitators of dialogue and inquiry, interpreters of complex human behavior, and ethical stewards who mediate the uncertainties that AI often amplifies rather than resolves. Their relational presence, expressed through empathy, intuition, cultural awareness, and ethical discernment, forms the bedrock of student trust and meaningful engagement. No algorithm, regardless of sophistication, can replicate the commitment of a teacher who perceives potential where data sees deficits, who recognizes context where models see anomalies, and who affirms humanity where systems see patterns.

Students, in turn, emerge not as passive recipients of information but as active agents navigating a world shaped by intelligent systems. The educational task is no longer simply to transmit knowledge but to cultivate the capacities students need to participate ethically in this world: critical thinking, self-direction, collaboration, digital literacy, and the moral imagination to know when to embrace technology and when to challenge it. AI does not diminish the importance of students' intellectual work; rather, it demands more sophisticated forms of reasoning and reflection, elevating the cognitive and ethical expectations placed upon learners.

These changing roles also operate within a wider framework that requires deliberate planning. Classrooms enriched by AI must remain unmistakably human spaces where identity is honored, relationships are nurtured, and community is built through shared inquiry. School cultures must uphold transparency, guard privacy, confront bias, and ensure that technological adoption strengthens rather than constrains teacher autonomy. In such cultures, AI is an ally in the pursuit of equity and flourishing. Without such cultures, even the most promising tools risk magnifying inequities, narrowing learning, or eroding trust.

The future of education with artificial intelligence will be shaped not by machines, but by the wisdom of the humans who guide them. The responsibility, and opportunity, before teachers, students, and educational leaders is to shape a future in which technology amplifies the virtues of teaching rather than diminishes them. If we commit to a vision grounded in humanity, ethics, and relational understanding, AI can help build classrooms where creativity thrives, inquiry deepens, and every student is seen not as a data point, but as a person with inherent worth and limitless potential.

The challenge of this new era is significant, but so is the promise. By establishing roles with precision and thoughtful consideration, it is possible to foster an educational setting in which human and machine intelligence effectively collaborate. This approach supports the fundamental objectives of education: cultivating wisdom, developing character, and equipping learners to contribute to a fair and compassionate society. As artificial intelligence becomes more prevalent in classrooms, it is essential that we remain vigilant in preserving the core values of empathy, critical thinking, and relational trust that define meaningful learning. The future of teaching and learning with AI is not a future without teachers or students; rather, it is a future in which both stand stronger, with teachers as guides, mentors, and ethical stewards, and students as active agents and thoughtful citizens, guided by the timeless truth that learning is, above all, a profoundly human endeavor. When technology is purposefully used to foster connection, promote equity, and advance human development, education can continue to empower both individuals and communities. This approach prepares learners to navigate and thrive amid the possibilities and challenges presented by a world increasingly shaped by artificial intelligence.

As roles evolve, so must the systems that support them. Teachers cannot be asked to serve as ethical stewards of AI without clear governance, privacy protections, and guidance on academic integrity, procurement, and accountability. Chapter 11 turns from classroom roles to the structural conditions that make responsible AI use possible at scale, showing how policy and equity commitments can guide adoption, so innovation strengthens public trust rather than undermines it.

Chapter 11:
Policy, Equity, and the Path Forward

The previous chapter argued that AI changes roles: teachers become designers, coaches, and ethical stewards, and students must learn to use powerful tools without surrendering their own thinking. Those shifts cannot be sustained by classroom practice alone. They require clear policies that protect student dignity, reduce inequity, and keep high-stakes decisions accountable to human judgment. This chapter focuses on the system layer, how schools, districts, and states can govern AI responsibly, so innovation strengthens, rather than weakens, equity and public trust.

The rapid integration of artificial intelligence (AI) in educational systems marks a turning point in how societies conceive of learning, governance, and the public mission of schooling. While previous chapters have centered on the irreplaceable humanity of teachers and students, this chapter turns to the broader structural question: how must educational policy evolve to ensure that AI enhances, rather than undermines, the principles of equity, justice, and democratic participation? The answers to this question will determine not only how AI is used, but who benefits from its use and, equally important, who may be harmed by it.

Policy goes beyond technical or administrative tools; it is a moral statement embodying shared values. Historically, major shifts in educational technology from radio instruction to computing have been shaped as much by political priorities as by pedagogical aspirations (Cuban, 1986). AI heightens these dynamics because it not only reorganizes instructional practice but also reconfigures power: power over information, decision-making, identity formation, and future opportunity. Without explicit attention to ethics and equity, AI implementation risks reinforcing the very inequities education was created to counteract. Scholars such as Benjamin (2019) and Noble (2018) warn that digital systems often mirror historical injustices embedded in data, disproportionately affecting marginalized communities. The introduction of AI in schools brings these concerns into sharper focus.

As introduced in Chapter 2, responsible AI use in education rests on six pillars: Transparency, Accountability, Equity, Privacy, Governance, and Human Judgment. Figure 11.1 gathers those commitments in one place. The sections that follow apply them through the chapter's themes, human-centered innovation, equity risks and responsibilities, fairness and recourse, policy pathways, and a justice-oriented vision, before closing with an implementation roadmap that turns principles into action.

AI holds genuine potential for reducing barriers to learning by personalizing instruction, identifying unmet needs, supporting multilingual students, and offering tools to assist learners with disabilities. These possibilities raise an essential question: How do we design policies that unlock AI's benefits while mitigating its risks? The answer begins with recognizing that equity cannot be an afterthought. It must be a guiding principle that shapes policy from the outset, determining what technologies are adopted, how and when they are used, and who has a voice in those decisions.

The path forward requires moving beyond simple narratives of innovation. AI is often introduced in schools with promises of efficiency, predictive power, or competitive advantage. Yet such narratives can obscure the more foundational purposes of education, such as preparing informed citizens, nurturing human potential, and promoting social good. As Pasquale (2020) argues, societies must resist the tendency to allow automation to erode human judgment, especially in domains where ethics, relationships, and human dignity are central. Policy, therefore, must serve as a safeguard that ensures AI aligns with rather than overrides public values. Policy must do two things at once: protect students and guide innovation. In the context of AI, that means building frameworks that promote equity, transparency, privacy, and professional autonomy while ensuring that decision-making authority remains with humans rather than algorithms.

Figure 11.1

Human-Centered Policy Principles for AI Governance in Education

Figure 11.1 shows how the pillars function together in practice. Governance defines who decides and how decisions are reviewed; transparency and accountability make those decisions explainable and answerable; equity and privacy protect students from disproportionate harm; and human judgment ensures that high-stakes educational choices remain rooted in professional and ethical discernment rather than automated output.

The road ahead is complex, but it is navigable. By foregrounding equity and ethical responsibility, policymakers and educators can shape

AI's role in schools to deepen learning, strengthen relationships, and uphold the promise of education as a public good. This chapter offers a pathway for doing so, one grounded not in technological determinism, but in the enduring values that define a just and humane society.

Human-Centered Policy for Responsible AI Innovation

Policy in education has always been a negotiation between competing values: efficiency and equity, autonomy and accountability, innovation and stability. The emergence of AI intensifies these tensions. Without careful policy, schools may use AI that values efficiency above human connection, surveillance over trust, and standardization instead of meaningful learning (OECD, 2020). The real challenge is aligning AI with education's broader goals, not merely regulating it. International policy guidance has likewise emphasized that AI in education must be governed through frameworks that support inclusion, equity, transparency, and institutional responsibility (UNESCO, 2021). A trustworthy approach to educational AI requires transparency, accountability, and safeguards that protect learners and educators from harm (OECD, 2020).

A human-centered policy approach begins with transparency. Schools must explain what AI tools are being used, why they are being used, what data they collect, and how decisions are made. Transparency builds trust among teachers, students, and families and makes ethical oversight possible. Accountability is equally essential. AI systems may provide recommendations, but responsibility for educational decisions must remain with humans. Policies must explicitly affirm that teachers, not algorithms, hold final decision-making authority. This protects professional autonomy and ensures that contextual knowledge informs educational choices. As Pasquale (2020) notes, accountability in the age of automation requires "defending human expertise" by preventing the outsourcing of moral judgment to machine-driven systems.

Policy must also address the governance of student data. AI systems require access to information that contains attendance, behavior, academic performance, and often more sensitive data to function effectively. Without strict data protections, students can be exposed to privacy violations, stigmatizing profiles, or long-term consequences resulting from algorithmic misinterpretations. Policies that emphasize data minimization, clear consent processes, and strict limits on third-party access can protect students from harm while preserving the benefits of responsible AI use.

At the heart of human-centered innovation is the recognition that technological adoption must never outpace ethical reflection. Emerging state guidance also shows that human-centered AI implementation in schools requires practical expectations for local oversight, staff support, and responsible instructional use (Washington Office of Superintendent of Public Instruction, 2024). This expectation is consistent with broader international ethical commitments that frame AI governance around human rights, dignity, fairness, and accountability (UNESCO, 2021b). Schools need policies that allow for continuous review of AI systems, including audits of algorithmic bias, mechanisms for reporting harm, and structures for revising or discontinuing AI tools that fail to meet ethical standards. Such policies ensure that innovation remains iterative, reflective, and guided by human values.

In these ways, policy is both a protection and an opportunity: it safeguards the dignity of students and teachers while enabling innovations that deepen learning, expand access, and restore time for meaningful work. When aligned with human-centered principles, policy becomes the anchor that prevents AI from drifting toward practices that undermine equity or erode trust.

AI and Educational Equity: Risks, Realities, and Responsibilities

Artificial intelligence enters educational systems that are already marked by deep and persistent inequities. As AI functions within existing societal frameworks, its implementation in education prompts critical inquiry into whether this technology will serve to challenge systemic inequities or reinforce them further. Schools vary greatly in funding, teacher quality, advanced classes, and community resources. These inequities are not incidental; they reflect the broader social histories of segregation, poverty, and racialized opportunity gaps that define the educational environment in many communities (Benjamin, 2019; Noble, 2018). When AI tools are layered upon systems with these entrenched disparities, they inevitably absorb their contours. This challenge becomes especially pronounced when predictive analytics rely on datasets shaped by decades of uneven opportunities. A student flagged as "at risk" by an algorithm may simply be the product of a system that has historically failed to support similar students, thereby perpetuating cycles of disadvantage rather than illuminating new solutions (Crawford, 2021).

Bias within AI systems surfaces in subtle yet consequential ways, often in forms invisible to educators and policymakers. Automated essay scoring systems may undervalue linguistic diversity; behavior-tracking algorithms may misinterpret cultural communication styles as signs of disengagement; and facial recognition technologies may demonstrate higher error rates for students with darker skin tones (Benjamin, 2019; Noble, 2018). Even when the data appear objective, the metrics upon which AI relies can reproduce historical disparities. Student disciplinary records, for example, reflect longstanding patterns of disproportionate punishment targeting people of color and non-English speakers. When algorithms use such records to make predictions, these inequities become embedded in the logic of the system. This process risks transforming historical injustice into digital inevitability, reinforcing the very patterns AI claims to solve. Rosenthal and Jacobson (1968) show that expectations, both human and algorithmic, influence outcomes and can limit student potential.

Equity concerns also extend to resource distribution and access, revealing another dimension of technological inequality. AI-enhanced learning environments require consistent internet connectivity, updated devices, and responsive technical support, which are conditions not uniformly available across schools. Students in under-resourced communities may be unable to benefit from personalized learning systems or generative AI tools that support creative expression or multilingual learning. In such contexts, AI becomes not a bridge to opportunity but another fault line dividing those with access from those without. Disparities also affect educators; institutions with limited resources frequently lack sufficient means to offer professional development that equips teachers with the knowledge to evaluate and responsibly utilize AI tools. As Williamson and Piattoeva (2023) caution, the increasing datafication of education risks concentrating power in the hands of vendors and policymakers unless teachers and school administrators are equipped to engage critically with the technologies that shape their work.

Surveillance issues add complexity to equity. AI-driven monitoring tools that range from keystroke tracking to facial recognition are often introduced in the name of safety or efficiency, yet their impact is rarely neutral. Students who already experience disproportionate disciplinary scrutiny are most likely to be negatively affected by heightened monitoring. For these learners, AI intensifies the sense of being watched, evaluated, or controlled, potentially undermining their trust in school environments. Privacy concerns make these problems worse: data about academic performance, behavior, or emotions can be kept long-term, shared with others, or used to classify students for the duration of their education (Crawford, 2021). Students with the least institutional power bear the greatest consequences when data governance fails.

Despite these risks, the story of AI and equity need not be one of harm. AI holds genuine potential to expand access, personalize learning, and support students whose needs have historically been overlooked. Intelligent tutoring systems can help close foundational skill gaps. Assistive technologies can empower students with disabilities. Language models can offer scaffolding that supports English learners in accessing rigorous content. Predictive systems, when carefully designed and ethically interpreted, can help teachers identify students in need of support earlier than would otherwise be possible. Yet these benefits materialize only when equity is intentionally placed at the center of AI design and implementation.

For AI to contribute to justice in education, policymakers and educators must adopt a posture of vigilance and responsibility. This requires auditing systems for bias, ensuring equitable access to infrastructure and training, elevating and embracing teacher and community voices in decision-making processes, and developing transparent data governance practices that protect student privacy and dignity. Pasquale's (2020) insistence on defending human expertise offers an important guidepost: equity cannot be automated, and the moral responsibilities of education cannot be delegated to algorithms. Justice in an AI-rich future will depend on educators and leaders who understand that technology must serve human flourishing, not the reverse.

AI will not automatically create a more equitable education system. Equity is a human achievement, not a technological one. It requires intentional policy, ethical stewardship, and a clear commitment to seeing students not as data points but as individuals shaped by history, identity, and possibility. When equity becomes the foundation rather than the afterthought of AI implementation, AI can support a more inclusive and humane path forward, one that aligns technological advancement with the enduring moral commitments of public education.

Ensuring Equity and Fairness in Educational AI Use

Guaranteeing fairness in AI systems extends beyond technical considerations; it is both a moral and educational responsibility. When artificial intelligence is implemented within educational settings, ensuring fairness becomes integral to upholding the institution's legitimacy. Schools are obligated to uphold equity, affirm student dignity, and create conditions in which all learners can thrive. AI systems, by contrast, operate according to statistical patterns, computational efficiencies, and datasets shaped by existing social structures. Bridging the gap between these differing logics requires intentional design, vigilant oversight, and a commitment to examining not only what AI systems do, but what they assume about the students they serve.

Fairness begins with acknowledging that AI systems are never neutral. Each model reflects choices about which data to include which metrics to prioritize, and which outcomes to predict. These choices mirror the values, explicit and implicit, held by developers, vendors, policymakers, and institutions. Even when AI appears to perform effectively at scale, it may do so at the cost of misreading or misrepresenting certain student groups whose lived experiences fall outside the boundaries of trained patterns. As scholars such as Noble (2018) and Benjamin (2019) have shown, seemingly objective systems often reinforce inequalities when they are built atop data shaped by racial, linguistic, and socioeconomic disparities. Ensuring fairness, therefore, requires educators and policymakers to interrogate not only the outputs of AI systems but the assumptions embedded within their design.

One essential component of fairness involves examining how AI makes predictions. Algorithms trained on historical data may reproduce patterns of discrimination without recognizing the context behind them. A model that predicts future performance based on past disciplinary referrals, for example, may simply encode the biases of systems known to disproportionately target students of color and non-English speaking students. Ensuring fairness requires educational institutions to question whether the patterns AI identifies reflect student potential or systemic failure. When predictive analytics are used to assign interventions, categorize learners, or shape opportunities, the stakes are especially high. When an algorithm identifies a student, it applies more than a neutral designation; it assigns the student to a pathway that may affect expectations, opportunities, and self-perception (Rosenthal & Jacobson, 1968). Fairness demands that educators treat such predictions not as conclusions but as prompts for deeper investigation.

Fairness also rests on transparency. Without clear and comprehensible explanations of how AI systems operate, students, teachers, and families cannot assess whether recommendations are fair or flawed. Strategies such as opening datasets, documenting model assumptions, and describing decision pathways in language accessible to non-specialists help demystify AI systems and enable stakeholders to question their outputs. Transparency transforms AI from a black box into a starting point for dialogue. It allows teachers to compare algorithmic recommendations with contextual knowledge and empowers families to voice concerns when systems appear to misinterpret their children's needs. Crawford (2021) asserts that transparency represents not only a best practice, but also an ethical responsibility when technology impacts individuals and society.

Yet transparency alone does not ensure fairness. AI systems require continuous auditing, which is the formal process for examining how models behave across different student populations. Audits must evaluate whether systems produce disparate impacts, misinterpret linguistic or cultural expressions, or systematically recommend less supportive interventions for certain groups. These audits should not be occasional or superficial, but ongoing and embedded within policy and governance practices. They require the involvement of educators, researchers, and community members, not only technologists, because fairness in education cannot be evaluated through algorithms alone. It must be measured through human insight, relational understanding, and ethical reflection.

Fairness also depends on meaningful options for recourse. Students and families need clear pathways for challenging algorithmic decisions, correcting errors, or requesting human review. Teachers must retain the authority to override AI recommendations when contextual expertise reveals a more nuanced understanding of the situation. Pasquale (2020) emphasizes that systems designed to "defend human expertise" must institutionalize mechanisms that return decision-making power to those who understand the human context best. Recourse ensures that AI systems do not become silent arbiters of opportunity but remain tools subject to human interpretation and revision.

Finally, fairness requires a fundamental reorientation of purpose. AI should not be used to sort, rank, or pathologize students; it should be used to support them. When designed and governed with equity in mind, AI can help identify unmet needs, personalize instruction, assist multilingual learners, elevate student voices, and reveal patterns that might otherwise remain unseen. But such possibilities emerge only when fairness is placed at the center of AI development and implementation. It is not incidental to the process; it is the process.

Policy does not hinder progress; it provides the foundation that makes progress possible.

Policy Pathways for Ethical and Equitable AI Adoption

Establishing a responsible framework for artificial intelligence in education demands policies that are both forward-thinking and effective. Such policies must acknowledge the significant ethical implications of AI while fostering purposeful innovation. Ethical AI adoption does not unfold automatically; it must be guided by governance structures that anticipate risks, uphold human dignity, and ensure that technology serves pedagogical rather than commercial or administrative priorities. Policy does not hinder progress; it provides the foundation that makes progress possible. Without clear policies, AI integration in schools may become fragmented or vendor-driven rather than guided by educational priorities.

An essential starting point is the establishment of clear governance frameworks. Such frameworks articulate the principles and processes that guide AI adoption, including how tools are selected, how data are managed, and how decisions are evaluated. Governance frameworks must be rooted in transparency, describing in accessible language what AI systems do, how they function, and what assumptions they embed. Policies that clearly outline data flows in terms of what is collected, for what purpose, for how long, and with whom it is shared, and allow students, families, and educators to understand the technology's role and limitations (Crawford, 2021). Transparency fosters trust, and trust is the precondition for any successful integration of AI in schools.

Beyond governance structures, ethical AI adoption requires rigorous oversight mechanisms. Oversight keeps technology under ongoing review, not just deployment. Policies should mandate regular audits for algorithmic bias, require human review of automated decisions, and specify procedures for evaluating whether AI tools are achieving intended outcomes. Oversight must also be dynamic and ongoing, recognizing that AI systems evolve over time as data patterns shift and new uses emerge. It must also be participatory, involving teachers, students, families, and community members, not only technical experts or administrators. As Pasquale (2020) emphasizes, oversight rooted in democratic participation helps ensure that AI systems reflect community values rather than narrow technological or economic priorities.

Another critical policy pathway involves protecting and reinforcing teacher autonomy. As earlier chapters established, AI can support but never replace the contextual expertise teachers bring to their work. Policy must therefore assert explicitly that educators are the ultimate decision-makers in matters involving student learning and welfare. AI-generated insights should be framed as advisory, not authoritative, ensuring that teachers retain the right, and responsibility, to interpret, question, or reject algorithmic recommendations based on their professional judgment. Policies that safeguard teacher autonomy do more than protect professional identity; they ensure that educational decisions remain grounded in human understanding rather than automated pattern recognition (Ferreyra & Hatcher-Mbu, 2026).

Equally important is the development of robust data-privacy protections. Students entrust schools with sensitive information, and AI tools often require access to data that reveal not only academic performance but patterns of behavior, attendance, and emotional states. Policy must ensure that such data are collected minimally, stored securely, and shared only when essential. Schools must resist pressures to expand data collection for the sake of algorithmic optimization, which risks normalizing forms of surveillance that compromise student dignity. Policies that prioritize data minimization, informed consent, and strict limitations on third-party access uphold the ethical obligation to protect students from harm (Williamson & Piattoeva, 2023).

To foster truly ethical AI adoption, policies must also prioritize equitable access to technology and professional learning. Without an adequate infrastructure that includes reliable internet, sufficient devices, and continuous technical support, students in under-resourced schools cannot benefit from AI-enhanced learning. Similarly, without ongoing professional development, teachers cannot engage critically with AI systems, interpret data responsibly, or recognize algorithmic bias. Policies that invest in infrastructure and teacher learning are not peripheral; they are equity measures that ensure AI does not become another mechanism through which social disparities deepen (Noble, 2018; Benjamin, 2019).

Finally, ethical AI adoption requires spaces for reflection and redress. Policy must create avenues for students and families to question automated decisions, contest inaccuracies, or request human evaluation. Likewise, teachers must have mechanisms for reporting concerns, recommending changes, or discontinuing tools that produce harmful or inequitable outcomes. These spaces for reflection are essential because fairness cannot be automated; it emerges from human dialogue, ethical deliberation, and continuous refinement.

In essence, policy pathways for ethical AI adoption must be grounded in a commitment to humanity. They must ensure that technology remains a tool in the service of educational goals rather than a force that dictates them. Policies emphasizing transparency, oversight, autonomy, privacy, equity, and reflective practice establish a robust framework for integrating AI in a manner that upholds the ethical principles of education.

A Justice-Oriented Vision for AI in Education

A justice-oriented vision for AI in education begins with a foundational recognition that technology cannot transform schools unless the systems into which it is introduced are simultaneously transformed. AI possesses immense potential to support learning, but its value is realized only when schools commit to addressing inequity not as a technical problem but as an ethical mandate. Ensuring justice in the integration of AI necessitates that its adoption is informed by respect for human dignity, democratic engagement, and recognition of each learner's potential. Absent this foundation, AI may perpetuate existing systems of privilege instead of helping to address them.

This vision regards AI not as a final objective, but rather as an instrument for broadening human potential. This means prioritizing applications that deepen access to meaningful learning experiences that support multilingual learners, enable students with disabilities to participate more fully, illuminate overlooked patterns in academic or social-emotional needs, and help teachers respond with greater nuance and care. Yet even as AI enhances these capacities, educators must remain vigilant about the moral choices embedded in every technological design. Justice requires asking not only whether AI can improve efficiency but whether it helps cultivate a more inclusive and equitable educational environment.

Central to this vision is a commitment to equity as a proactive principle rather than a reactive correction. Historical patterns of racism, linguistic marginalization, and socioeconomic inequity shape the data on which AI systems are built (Benjamin, 2019; Noble, 2018). A justice-oriented policy acknowledges this reality and insists that AI systems undergo rigorous scrutiny before they are deployed. This includes examining whether training data reflect harmful biases, whether the model's assumptions align with human values, and whether the system's potential harms will be disproportionately borne by marginalized communities. Crawford (2021) reminds us that data are never neutral; they are cultural artifacts that reflect the values, exclusions, and power dynamics of the societies that generate them. Justice requires that these realities be confronted directly rather than obscured behind technical abstraction.

A justice-oriented vision also demands participatory governance. Decisions about AI adoption cannot reside solely with administrators, vendors, or policymakers; they must involve the voices of those most affected: students, families, and educators. Particularly important are the perspectives of communities historically marginalized by educational systems, whose insights can illuminate risks and possibilities that might otherwise remain invisible. Pasquale (2020) emphasizes that democratic oversight is essential when automated systems shape public life. In education, this involves designing systems where community involvement is a key part of making decisions, not just offering advice.

Another defining feature of justice-oriented AI is an unwavering commitment to human flourishing. Schools must reject applications of AI that narrow learning to what is easily measured or that reduce students to predictive profiles. Instead, they should prioritize technologies that foster creativity, support dialogue, nurture curiosity, and strengthen relationships. This requires resisting pressures to use AI for high-stakes decision-making such as grading, placement, and discipline, where the risks of misinterpretation or systemic bias are greatest. Justice demands that the most consequential educational decisions remain firmly in human hands, grounded in contextual understanding and moral reasoning (Holmes et al., 2019).

Justice also involves sustained investment in the schools and communities that have historically received the least. AI can widen the digital divide if access to devices, infrastructure, training, and support remains uneven. A justice-oriented policy insists that funding for AI initiatives be tied to equitable distribution, ensuring that under-resourced schools have the tools and professional learning necessary to implement AI responsibly. Equity cannot be achieved through technology alone; it requires structural commitments to fair funding, community partnerships, and ongoing support for teachers' professional growth (Williamson & Piattoeva, 2023).

A justice-oriented approach to AI in education is inherently adaptive. It acknowledges the evolving nature of technology, alongside changes in cultural norms, institutional goals, and societal expectations. Achieving justice necessitates ongoing evaluation, regular policy review, adjustment of assumptions, and responsiveness to new evidence. This approach requires humility, recognizing that unexpected outcomes may arise even from carefully designed systems and emphasizing the importance of sustained ethical oversight.

Justice should be understood as an ongoing process: an informed, principled engagement with AI, prioritizing responsibility and the value of human connection. As educational contexts shift and AI capabilities advance, this approach insists on continuous dialogue among stakeholders: educators, students, families, and policymakers to ensure that practices remain relevant and equitable. Through this collective vigilance and willingness to learn from both successes and setbacks, schools can foster an environment where technology supports, rather than undermines, the moral and relational heart of education.

By orienting AI integration toward justice, schools affirm that technology must never overshadow the humanity it is intended to support. Such a vision ensures that AI is used not to replicate the inequities of the past but to help build a future rooted in dignity, opportunity, and shared flourishing. It envisions education not as a site of algorithmic sorting, but as a community where every learner is seen, valued, and empowered.

Implementation Roadmap

Moving from principles to practice requires more than enthusiasm for innovation; it requires a disciplined implementation process that protects students while giving educators room to learn. The roadmap below translates this chapter's commitments of transparency, accountability, equity, privacy, governance, and human judgment into a practical sequence of actions. It is designed to help schools and districts avoid fragmented adoption, strengthen public trust, and ensure that AI use remains consistent with the moral purposes of education. In practice, leaders should begin by clarifying purpose by identifying the most urgent problems they are trying to solve, such as workload, access, or feedback cycles. They should define the specific instructional use cases for AI and establish success measures that include equity indicators. From there, districts should set guardrails before expanding access by adopting governance structures, privacy protections, and academic integrity expectations, and by communicating clear decision rights and responsibilities.

With these boundaries in place, schools can pilot with integrity through limited trials in representative settings, monitoring outcomes and unintended consequences and documenting the teacher practices required to keep human judgment central. Only after pilots generate trustworthy evidence should systems scale with support, pairing expansion with professional learning, family communication, and ongoing monitoring that includes regular check-ins for equity and privacy impacts. Finally, leaders should review and revise routinely by treating AI policy as living guidance that evolves as tools and risks change, and by using regular review cycles to retire tools that do not meet instructional, ethical, or governance standards.

Taken together, these steps emphasize coherence over speed. They help leaders treat AI adoption as a continuous improvement effort that is evaluated in real classrooms, shaped by educator and community feedback, and adjusted as risks and expectations evolve. With this implementation sequence in view, the chapter closes by returning to its central claim: policy and governance determine whether AI strengthens equity and public trust or amplifies existing harms.

Artificial intelligence in education offers both opportunities and challenges. The key concern is how its effects are shaped through policy, equity, governance, and ethics. Thoughtful frameworks can prevent AI from reinforcing inequities and limiting educator agency, while justice-driven policies can help improve learning outcomes and expand access. Policies should focus on human dignity, ensuring fairness, transparency, trust, and respect. Technology must be designed for democratic participation, not just convenience. Schools use policy to protect privacy, uphold teacher autonomy, and distribute AI benefits fairly across all communities.

Equity requires a proactive approach. Artificial intelligence can reveal patterns, assist multilingual learners, increase accessibility for students with disabilities, and generate actionable insights that empower educators to respond more effectively. However, these advantages are only attainable when institutions address systemic inequities in data, scrutinize algorithmic bias, and ensure AI tools do not perpetuate surveillance or control. A justice-oriented perspective prioritizes technology to support students, guided by human judgment in all automated processes.

Progress requires ongoing collaboration among policymakers, educators, families, technologists, and communities. The trajectory of AI in education remains unsettled and will be shaped by ethical values, policy decisions, governance frameworks, and educational aims. Policies grounded in humanity support equity; innovation guided by equity builds trust; and trust allows AI to broaden access and strengthen core educational relationships. AI's promise will be realized only if education remains focused on its essential mission: fostering wisdom, promoting human development, and advancing social justice. Although the path ahead is complex, it is navigable. With clarity, determination, and a strong ethical foundation, educational institutions can build a future in which technology strengthens teaching and learning.

Conclusion:
Teaching Humanity in the Age of Intelligence

Education has always evolved alongside technology, from the printing press to the internet, and each new tool has reshaped how we think, teach, and learn. Yet artificial intelligence (AI) marks a turning point unlike any that came before. Earlier innovations extended human reach by distributing information more widely or making tasks faster. AI goes further. It does not merely store knowledge; it interprets patterns, generates language, and performs forms of cognitive work that increasingly influence how people communicate, create, and solve problems. Education is entering a new era in which teachers and students work alongside intelligent systems that can generate ideas, simulate understanding, personalize pathways, and respond in real time. The question before us is not whether AI belongs in education, but whether we will shape it to serve human flourishing rather than allow it to diminish the very capacities schooling exists to cultivate.

Throughout this book, we have seen AI's potential to relieve teacher workloads, personalize learning at scale, offer continuous feedback, deepen student inquiry, and help educators analyze learning with greater precision. Used wisely, AI can restore time and creativity to the classroom, giving teachers greater freedom to nurture curiosity, empathy, and moral imagination. It can help educators see patterns that would otherwise remain invisible, create more accessible materials for multilingual learners, and support students as they draft, revise, and persist. At the same time, we have acknowledged a parallel truth: tools designed to empower education can also erode human judgment if used without reflection. Overreliance on automation may narrow professional discretion, weaken students' ability to think independently, embed bias and inequality within opaque systems, and replace relationships with recommendations. AI can either amplify the best of education or reproduce its deepest flaws at scale. The outcome depends not on the technology itself, but on the humans who guide it.

To navigate this moment, we must balance automation with autonomy, data with discernment, and efficiency with empathy. That balance depends on three commitments. First, teachers and students must remain the final decision-makers, using AI as a partner for drafting and analysis while retaining responsibility for judgment, meaning, and ethics. Second, schools must adopt clear principles for transparency, privacy, and fairness, so that data use is bounded, explainable, and accountable to the communities it affects. Third, learners must be taught how to question, critique, and co-create with AI in ways that empower them to verify claims, document process, and use assistance without surrendering authorship or integrity. If we hold these commitments, technology can extend human potential rather than replace it.

The teacher in the age of AI is not primarily a transmitter of information; AI can already perform much of that function. But while artificial intelligence can facilitate the delivery of knowledge, educators cultivate wisdom by helping students consider both the practical uses of information and the ethical implications of their actions. Teachers create the conditions for belonging, perseverance, and intellectual courage. They notice what data cannot capture, interpret what algorithms cannot understand, and decide what is just when a recommendation may be efficient but not humane. As machines become more capable, the teacher's judgment, care, and wisdom become even more important. This truth should anchor education in every future we imagine.

For students, AI brings both opportunity and responsibility. Students can learn at their own pace, access wider fields of knowledge, and create beyond traditional limits, but they must also learn to discern truth from illusion, collaborate ethically, and remain grounded in humanity amid automation. They will need to understand how AI systems can be persuasive and wrong, how bias can shape outputs, and how convenience can become a subtle shortcut that weakens learning. The next generation must be prepared not only to use AI, but to guide its evolution with integrity and compassion.

AI in schools must therefore be used intentionally. Teachers should use it to enrich learning, students must engage with it respectfully and transparently, leaders must prioritize fairness and trust, and policymakers should treat it as a means to deepen understanding rather than merely increase efficiency. While AI can accelerate certain forms of learning, meaning is still made through human interaction, dialogue, struggle, mentorship, and the careful formation of values. Used well, AI can widen access to practice, feedback, and explanation; used poorly, it can shrink learning into compliance and trade wonder for convenience. Our task is to choose the first path with courage and clarity, so every new capability serves a deeper purpose: helping young people think more clearly, care more deeply, and imagine more boldly.

This book is not ultimately about technology. It is about humanity in an age of shared intelligence. The goal of AI in education is not replacement or automation, but the strengthening of human thinking, empathy, and creativity. When we remember this, we keep faith with the true work of schooling: to help students become wise, capable, and compassionate people. When we forget it, we risk gaining speed and losing direction. The future of education must be defined by genuine humanity, supported by tools, guided by wisdom, and anchored in dignity. If we lead with purpose, AI can become not a shortcut around learning, but a lantern within it, illuminating deeper understanding and calling forth the best in who we are.

AI can open doors to knowledge, but it is curiosity, courage, and conscience that teach us how to walk through them.

References:

Al-Azawei, A., Serenelli, F., & Lundqvist, K. (2016). Universal design for learning (UDL): A content analysis of peer-reviewed journal papers from 2012 to 2015. *Journal of the Scholarship of Teaching and Learning, 16*(3), 39–56.

Anderson, L. W., & Krathwohl, D. R. (Eds.). (2001). *A taxonomy for learning, teaching, and assessing: A revision of Bloom's taxonomy of educational objectives.* Longman.

Au, W. (2016). *Meritocracy 2.0: High-stakes testing, accountability, and education reform.* Routledge.

Azevedo, R. (2015). Defining and measuring self-regulated learning. In J. A. Greene & W. A. Sandoval (Eds.), *Handbook of cognition and assessment* (pp. 467–487). Wiley.

Benedikt, L., Joshi, C., Nolan, L., Henstra-Hill, R., Shaw, L., & Hook, S. (2020). Human-in-the-loop AI in government: A case study. In *Proceedings of the 25th International Conference on Intelligent User Interfaces* (pp. 488–497). ACM.

Benjamin, R. (2019). *Race after technology: Abolitionist tools for the new Jim code.* Polity.

Berners-Lee, T. (1999). *Weaving the web: The original design and ultimate destiny of the World Wide Web.* HarperCollins.

Biggs, J., & Collis, K. (1982). *Evaluating the quality of learning: The SOLO taxonomy (structure of the observed learning outcome).* Academic Press.

Bjorklund, D. F., & Causey, K. B. (2018). *Children's thinking: Cognitive development and individual differences* (6th ed.). Sage.

Black, P., & Wiliam, D. (1998). Inside the black box: Raising standards through classroom assessment. *Phi Delta Kappan, 80*(2), 139–148.

Bloom, B. S. (1956). *Taxonomy of educational objectives: The classification of educational goals.* Longmans, Green.

Bloom, B. S. (1984). The 2-sigma problem: The search for methods of group instruction as effective as one-to-one tutoring. *Educational Researcher, 13*(6), 4–16.

Bransford, J. D., Brown, A. L., & Cocking, R. R. (Eds.). (2000). *How people learn: Brain, mind, experience, and school.* National Academy Press.

Brynjolfsson, E., & McAfee, A. (2017). *Machine, platform, crowd: Harnessing our digital future.* W. W. Norton.

Bulger, M. (2016). *Personalized learning: The conversations we're not having.* Data & Society Research Institute. https://datasociety.net/pubs/ecl/PersonalizedLearning_primer_2016.pdf

Burns, M. (2026, January 27). *What the research shows about generative AI in tutoring.* Brookings. https://www.brookings.edu/articles/what-the-research-shows-about-generative-ai-in-tutoring/

Case, R. (1992). *The mind's staircase: Exploring the conceptual underpinnings of children's thought and knowledge.* Erlbaum.

CAST. (n.d.). *Universal Design for Learning guidelines version 3.0.* https://udlguidelines.cast.org/

CERN. (1993). *The World Wide Web project.* https://info.cern.ch/hypertext/WWW/TheProject.html

Cerullo, M. (2023, June 2). Texas judge bans filings solely created by AI after ChatGPT made up case. *CBS News.*

https://www.cbsnews.com/news/texas-judge-bans-chatgpt-court-filing/

Chi, M. T. H. (2009). Active-constructive-interactive: A conceptual framework for differentiating learning activities. *Topics in Cognitive Science, 1*(1), 73–105.

Crawford, K. (2021). *Atlas of AI: Power, politics, and the planetary costs of artificial intelligence.* Yale University Press.

Creagh, S. (2025). Workload, work intensification and time poverty: Teachers and school leaders. *Educational Research, 67*(2), 260–280.

Cuban, L. (1986). *Teachers and machines: The classroom use of technology since 1920.* Teachers College Press.

Cuban, L. (2013). *Inside the black box of classroom practice: Change without reform in American education.* Harvard Education Press.

Cuban, L. (2018). *The flight of a butterfly or the path of a bullet? Using technology to transform teaching and learning.* Harvard Education Press.

Darling-Hammond, L., Flook, L., Cook-Harvey, C., Barron, B., & Osher, D. (2020). Implications for educational practice of the science of learning and development. *Applied Developmental Science, 24*(2), 97–140.

Darling-Hammond, L., Wilhoit, G., & Pittenger, L. (2014). *Developing a new generation of assessments.* Stanford Center for Opportunity Policy in Education. https://edpolicy.stanford.edu

Denning, P. J., & Martell, C. H. (2015). *Great principles of computing.* MIT Press.

Diamond, A. (2013). Executive functions. *Annual Review of Psychology, 64,* 135–168.

Dieker, L. A., Rodriguez, J. A., Lignugaris-Kraft, B., Hynes, M. C., & Hughes, C. E. (2014). The potential of simulated environments in teacher education: Current and future possibilities. *Teacher Education and Special Education, 37*(1), 21–33.

Dumont, H., Istance, D., & Benavides, F. (Eds.). (2010). *The nature of learning: Using research to inspire practice.* OECD Publishing. https://doi.org/10.1787/9789264086487-en

Education Week Research Center. (2022, April 19). *How teachers spend their time: A breakdown. Education Week.*

Eisenstein, E. (1979). *The printing press as an agent of change.* Cambridge University Press.

Ekizoğlu, M., & Demir, A. N. (2025). The role of AI assisted writing feedback in developing secondary students writing skills. *Discover Education, 4*, Article 454. https://doi.org/10.1007/s44217-025-00919-3

Ellington, A. J. (2003). The effects of non-CAS graphing calculators on student achievement and attitude: A meta-analysis. *School Science and Mathematics, 103*(1), 16–26.

Evans, P., & Martin, A. (2024). Cognitive load theory and its relationships with motivation: A self-determination theory perspective. *Educational Psychology Review, 36*, 7.

Ferreyra, F., & Hatcher-Mbu, B. (2026, January 21). *Beyond the algorithm: The case for human judgment in AI.* Development Gateway: An IREX Venture. https://developmentgateway.org/blog/beyond-the-algorithm-the-case-for-human-judgment-in-ai/

Fink, L. D. (2003). *Creating significant learning experiences: An integrated approach to designing college courses.* Jossey-Bass.

Flavell, J. H. (1985). *Cognitive development* (2nd ed.). Prentice-Hall.

Fullan, M., & Langworthy, M. (2014). *A rich seam: How new pedagogies find deep learning.* Pearson.

Gay, G. (2018). *Culturally responsive teaching: Theory, research, and practice* (3rd ed.). Teachers College Press.

Hattie, J. (2009). *Visible learning: A synthesis of over 800 meta-analyses relating to achievement.* Routledge.

Hattie, J., & Timperley, H. (2007). The power of feedback. *Review of Educational Research, 77*(1), 81–112.

Heritage, M. (2018). *Formative assessment in practice: A process of inquiry and action.* Harvard Education Press.

Higgins, S., Beauchamp, G., & Miller, D. (2007). Reviewing the literature on interactive whiteboards. *Learning, Media and Technology, 32*(3), 213–225.

Hollands, F. M., & Tirthali, D. (2014). *MOOCs: Expectations and reality.* Center for Benefit-Cost Studies of Education, Teachers College, Columbia University. https://cbcse.org/wordpress/wp-content/uploads/2014/05/MOOCs_Expectations_and_Reality.pdf

Holmes, W., Bialik, M., & Fadel, C. (2019). *Artificial intelligence in education: Promises and implications for teaching and learning.* Center for Curriculum Redesign. https://curriculumredesign.org/our-work/artificial-intelligence-in-education/

Ifrah, G. (2000). *The universal history of numbers: From prehistory to the invention of the computer.* Wiley.

International Society for Technology in Education. (1998). *National educational technology standards for teachers.* https://iste.org/standards

Keating, D. P. (2004). Cognitive and brain development. In R. M. Lerner & L. Steinberg (Eds.), *Handbook of adolescent psychology* (pp. 45–84). Wiley.

Koedinger, K. R., & Corbett, A. T. (2006). Cognitive tutors: Technology bringing learning science to the classroom. In R. K.

Sawyer (Ed.), *The Cambridge handbook of the learning sciences* (pp. 61–78). Cambridge University Press.

Kolb, D. A. (1984). *Experiential learning: Experience as the source of learning and development*. Prentice Hall.

Koretz, D. (2017). *The testing charade: Pretending to make schools better*. University of Chicago Press.

Kuhn, D. (2009). Do cognitive changes accompany developments in the adolescent brain? *Perspectives on Psychological Science, 4*(4), 342–351.

Ladson-Billings, G. (2021). *Culturally relevant pedagogy: Asking a different question*. Teachers College Press.

Lee, V. R., & Wilkerson, M. H. (2018). *Data use by middle and secondary students in the digital age: A status report and future prospects*. Commissioned paper for the National Academies of Sciences, Engineering, and Medicine, Board on Science Education, Committee on Science Investigations and Engineering Design for Grades 6–12. https://sites.nationalacademies.org/cs/groups/dbassesite/documents/webpage/dbasse_189500.pdf

Luckin, R., Holmes, W., Griffiths, M., & Forcier, L. B. (2016). *Intelligence unleashed: An argument for AI in education*. Pearson. https://www.pearson.com/content/dam/corporate/global/pearson-dot-com/files/innovation/Intelligence-Unleashed-Publication.pdf

Marzano, R. J., & Kendall, J. S. (2007). *The new taxonomy of educational objectives* (2nd ed.). Corwin Press.

Meyer, A., Rose, D. H., & Gordon, D. (2014). *Universal design for learning: Theory and practice*. CAST Professional Publishing.

Mobilio, V., & Guglielmini, G. (2026). Rethinking schooling in the age of AI: Equity, ethics, and the four pillars of transformation. *Frontiers of Digital Education, 3*(1), 7. https://doi.org/10.1007/s44366-026-0081-3

Montalvo, T. (2025, June 4). *Prompt engineering: A critical new skillset for 21st-century teachers. eSchool News.* https://www.eschoolnews.com/digital-learning/2025/06/04/prompt-engineering-a-critical-new-skillset-for-21st-century-teachers/

National Research Council. (2012). *Education for life and work: Developing transferable knowledge and skills in the 21st century.* National Academies Press.

Noble, S. U. (2018). *Algorithms of oppression: How search engines reinforce racism.* NYU Press.

OECD. (2020). *Trustworthy artificial intelligence (AI) in education: Promises and challenges.* OECD Publishing. https://doi.org/10.1787/a6c90fa9-en

OECD. (2025). *What should teachers teach and students learn in a future of powerful AI? OECD Education Spotlights, 20.* OECD Publishing. https://doi.org/10.1787/ca56c7d6-en

Ok, M. W., Rao, K., Bryant, B. R., & McDougall, D. (2022). Special education technology: Universal design for learning and inclusive practices. *Journal of Special Education Technology, 37*(2), 67–79.

Oughtred Society. (2013). *The Oughtred Society slide rule reference manual.* https://www.oughtred.org/books/OSSlideRuleReferenceManual revA.pdf

O'Neil, C. (2016). *Weapons of math destruction: How big data increases inequality and threatens democracy.* Crown.

Pane, J. F., Steiner, E. D., Baird, M. D., & Hamilton, L. S. (2015). *Continued progress: Promising evidence on personalized learning.* RAND Corporation.

Papert, S. (1980). *Mindstorms: Children, computers, and powerful ideas.* Basic Books.

Pappano, L. (2012, November 2). The year of the MOOC. *The New York Times.* https://www.nytimes.com/2012/11/04/education/edlife/massi ve-open-online-courses-are-multiplying-at-a-rapid-pace.html

Pasquale, F. (2020). *New laws of robotics: Defending human expertise in the age of AI.* Belknap Press.

Pellegrino, J. W., DiBello, L., & Goldman, S. (2016). A framework for conceptualizing and evaluating assessment systems. *Educational Measurement: Issues and Practice, 35*(1), 16–28.

Perkins, D. N., & Salomon, G. (1992). Transfer of learning. In T. Husén & T. N. Postlethwaite (Eds.), *The international encyclopedia of education* (2nd ed., pp. 425–441). Pergamon.

Petroski, H. (1990). *The pencil: A history of design and circumstance.* Knopf.

Piaget, J. (1952). *The origins of intelligence in children.* International Universities Press.

Piaget, J. (1970). *Science of education and the psychology of the child.* Orion Press.

Piaget, J., & Inhelder, B. (1969). *The psychology of the child.* Basic Books.

Pianta, R. C., Hamre, B. K., & Allen, J. P. (2020). Teacher–student relationships and engagement: Conceptualizing, measuring, and improving the capacity of classroom interactions. In R. E. Slavin (Ed.), *Handbook of research on student engagement* (2nd ed., pp. 365–386). Springer.

Popham, W. J. (2017). *Classroom assessment: What teachers need to know* (8th ed.). Pearson.

Postman, N. (1985). *Amusing ourselves to death: Public discourse in the age of show business*. Viking.

Reagle, J. (2010). *Good faith collaboration: The culture of Wikipedia*. MIT Press.

Robson, E. (2008). *Mathematics in ancient Iraq: A social history*. Princeton University Press.

Rogoff, B. (2003). *The cultural nature of human development*. Oxford University Press.

Rose, D. H., & Meyer, A. (2006). *A practical reader in universal design for learning*. Harvard Education Press.

Roscoe, R. D., & McNamara, D. S. (2013). Writing Pal: Feasibility of an intelligent writing strategy tutor in the high school classroom. *Journal of Educational Psychology, 105*(4), 1010–1025.

Rosenthal, R., & Jacobson, L. (1968). *Pygmalion in the classroom*. Holt, Rinehart & Winston.

Saettler, P. (2004). *The evolution of American educational technology* (2nd ed.). Information Age Publishing.

Sawyer, R. K. (Ed.). (2014). *The Cambridge handbook of the learning sciences* (2nd ed.). Cambridge University Press.

Schmidt, E., & Cohen, J. (2013). *The new digital age: Transforming nations, businesses, and our lives*. Knopf.

Selwyn, N. (2019). *Should robots replace teachers? AI and the future of education*. Polity.

Shepard, L. A., Diaz-Bilello, E., Penuel, W. R., & Marion, S. F. (2020). *Classroom assessment principles to support teaching and learning.*

Center for Assessment, Design, Research and Evaluation, University of Colorado Boulder.

Shi, Y., Yu, K., Dong, Y., & Chen, F. (2026). *Large language models in education: A systematic review of empirical applications, benefits, and challenges. Computers and Education: Artificial Intelligence, 10,* 100529. https://doi.org/10.1016/j.caeai.2025.100529

Shermis, M. (2020). Contrasting state-of-the-art automated scoring of essays: Analysis. *Journal of Writing Analytics, 4,* 216–243.

Shulman, L. S. (1987). Knowledge and teaching: Foundations of the new reform. *Harvard Educational Review, 57*(1), 1–22.

Shute, V. J. (2008). Focus on formative feedback. *Review of Educational Research, 78*(1), 153–189.

Shultz, Sharon. (2023). *Universal design for learning.* NEA. https://www.nea.org/professional-excellence/student-engagement/tools-tips/universal-design-learning-introduction

Siegler, R. S., Eisenberg, N., DeLoache, J., Saffran, J. R., & Gershoff, E. (2022). *How children develop* (6th ed.). Worth Publishers.

Siemens, G., & Long, P. (2011). Penetrating the fog: Analytics in learning and education. *EDUCAUSE Review, 46*(5), 30–40.

Suppes, P. (1966). The uses of computers in education. *Scientific American, 215*(3), 206–220.

Sweller, J., van Merriënboer, J. J. G., & Paas, F. (2019). Cognitive architecture and instructional design: 20 years later. *Educational Psychology Review, 31*(2), 261–292.

Tamez-Robledo, N. (2024, March 14). *We know how much planning time teachers get on average. Is it enough? EdSurge.*

TNTP. (2025, July 1). *Teachers' time use: A review of the literature* [Executive summary]. TNTP.

Tomlinson, C. A. (2014). *The differentiated classroom: Responding to the needs of all learners* (2nd ed.). ASCD.

Traxler, J. (2009). Learning in a mobile age. *International Journal of Mobile and Blended Learning, 1*(1), 1–12.

Uygun, D. (2024). Teachers' perspectives on artificial intelligence in education. *Advances in Mobile Learning Educational Research, 4*(1), 931–939. https://doi.org/10.25082/AMLER.2024.01.005

UNESCO. (2021). *AI and education: Guidance for policy-makers.* https://unesdoc.unesco.org/ark:/48223/pf0000376709

UNESCO. (2021). *Recommendation on the ethics of artificial intelligence.* https://unesdoc.unesco.org/ark:/48223/pf0000380455

UNESCO. (2023). *Guidance for generative AI in education and research.* https://unesdoc.unesco.org/ark:/48223/pf0000386693

UNESCO. (2024). *AI competency framework for teachers.* https://unesdoc.unesco.org/ark:/48223/pf0000391104

VanLehn, K. (2011). The relative effectiveness of human tutoring, intelligent tutoring systems, and other tutoring systems. *Educational Psychologist, 46*(4), 197–221.

Vygotsky, L. S. (1978). *Mind in society: The development of higher psychological processes.* Harvard University Press.

Walkington, C. (2013). Using adaptive learning technologies to personalize instruction: The role of student interests and identities. *Journal of Educational Psychology, 105*(4), 932–945.

Washington Office of Superintendent of Public Instruction. (2024). *Human-centered artificial intelligence in schools.* https://ospi.k12.wa.us/student-success/resources-subject-area/human-centered-artificial-intelligence-schools

Watters, A. (2015, March 12). A brief history of calculators in the classroom. *Hack Education.* https://hackeducation.com/2015/03/12/calculators

Wiliam, D. (2011). *Embedded formative assessment.* Solution Tree Press.

Williamson, B., & Piattoeva, N. (2023). *Education governance and datafication: The new politics of digital education.* Routledge.

Wing, J. (2006). Computational thinking. *Communications of the ACM, 49*(3), 33–35.

Woolf, B. P. (2021). *Building intelligent interactive tutors: Student-centered strategies for revolutionizing e-learning.* Morgan Kaufmann.

Wu, X., Xiao, L., Sun, Y., Zhang, J., Ma, T., & He, L. (2022). A survey of human-in-the-loop for machine learning. *Future Generation Computer Systems, 135,* 364–381.

Zenderland, L. (1998). *Measuring minds: Henry Herbert Goddard and the origins of American intelligence testing.* Cambridge University Press.

Zimmerman, B. J. (2002). Becoming a self-regulated learner: An overview. *Theory Into Practice, 41*(2), 64–70.

Appendix A:
Educator Decision Guide for
Responsible AI Use

Artificial intelligence can expand what teachers are able to design, differentiate, and manage, but its value depends on whether it is used in ways that preserve human judgment, protect student dignity, and support authentic learning. Throughout this book, AI is presented not as a replacement for teaching, but as a tool that must remain subordinate to professional expertise, ethical reasoning, and relational care. This appendix is designed to help educators make sound decisions about when AI use is appropriate, when it requires caution, and when it should not be used at all.

The purpose of this guide is practical. Teachers and school leaders are increasingly asked to make immediate choices about AI-assisted lesson planning, feedback, assessment, communication, intervention, and policy. Those choices are often made under pressure, with limited time and uneven institutional guidance. A clear decision process can help educators act with greater consistency and confidence while remaining aligned with the central principles of this book: human oversight, ethical responsibility, transparency, equity, and the preservation of meaningful student thinking.

The first question educators should ask is simple: What is the educational purpose of using AI here? If the answer is merely convenience, speed, or novelty, the use of AI should be reconsidered. If the answer is that AI may help improve access, reduce unnecessary administrative burden, support differentiation, or create more time for direct human engagement, the use may be justified. AI is most defensible when it strengthens teaching and learning without weakening student agency or replacing professional judgment.

The second question is whether the task requires human interpretation, ethical sensitivity, or contextual understanding. If the task depends heavily on those qualities, AI should not be allowed to operate independently. For example, teachers may use AI to draft lesson materials or summarize patterns in student work, but they should not allow AI to make final judgments about a student's character, intent, emotional condition, or academic integrity. Whenever the stakes are high, the human educator must remain the final decision maker.

The third question is whether the use of AI preserves authentic evidence of learning. If students are expected to demonstrate their own reasoning, analysis, voice, or problem-solving, then AI must not do the cognitive work for them. A useful standard is this: AI may support the learning process, but it should not replace the learner's intellectual contribution. Teachers should therefore distinguish between AI as scaffold and AI as substitute. When AI helps students clarify directions, organize ideas, revise language, or access content, it may support learning. When it generates the answer, argument, or final performance the student is supposed to produce, it undermines the integrity of the task.

A practical way to apply these principles is to sort possible uses of AI into three categories: appropriate use, cautionary use, and inappropriate use.

1. Appropriate use includes tasks in which AI supports efficiency, accessibility, or instructional responsiveness without displacing teacher judgment or student thinking. Examples include drafting lesson ideas, generating alternate reading levels, translating directions, creating practice questions, organizing notes, identifying broad trends in student errors, and providing early-stage writing suggestions that a student must still evaluate

and revise. In these cases, AI functions as an assistant rather than an authority.

2. Cautionary use includes tasks that may be appropriate only with strong oversight and clear boundaries. Examples include automated feedback on writing, AI-generated grading suggestions, predictive alerts about student risk, AI-supported parent communication, and tools that recommend interventions based on student data. These uses may save time or reveal patterns, but they also carry risks related to bias, overreliance, privacy, and misinterpretation. In these cases, educators should verify outputs, examine context, and ensure that no action is taken solely because a system recommended it.

3. Inappropriate use includes tasks in which AI replaces moral judgment, obscures authorship, weakens authentic assessment, or makes high-stakes determinations about students without human review. Examples include allowing AI to determine final grades without teacher oversight, using AI alone to infer motivation or intent, delegating disciplinary decisions to predictive systems, using AI to generate student work presented as original, or relying on AI detection tools as definitive proof of misconduct. In these cases, AI use is inconsistent with responsible educational practice because it shifts decisions away from the human relationships and contextual expertise on which fair schooling depends.

Educators can also use the following five-question screen before adopting any AI-supported task or tool.

*   First, does this use of AI improve learning, access, or instructional responsiveness in a way that matters? If the benefit is superficial, the tool is probably unnecessary.

- Second, does this use preserve teacher authority and professional judgment? If the teacher becomes a passive recipient of automated decisions, the use is not appropriate.

- Third, does this use protect authentic student thinking and preserve meaningful evidence of learning? If students can bypass the intended cognitive work, the task should be redesigned.

- Fourth, does this use protect student privacy, dignity, and fairness? If the tool depends on unclear data practices, opaque recommendations, or potentially biased outputs, caution is required.

- Fifth, could I clearly explain this use to students, families, and colleagues without embarrassment or defensiveness? If the answer is no, the use likely needs revision.

This guide is especially important in assessment. Teachers should be wary whenever AI makes a task easier to complete without requiring genuine understanding. If an assignment can be successfully performed by a chatbot with little or no student thinking, the problem is not merely student misuse; it may be weak assessment design. In such cases, teachers should redesign tasks to emphasize process, reasoning, reflection, oral explanation, revision history, contextual application, or in-class performance. Responsible AI use in assessment is inseparable from responsible assessment design.

This guide also applies to teacher preparation and leadership. Teachers, school administrators, and district decision makers should not ask only whether a tool works. They should ask what kind of professional practice the tool encourages. Does it cultivate reflective judgment or passive dependence? Does it strengthen equity or widen gaps? Does it create more space for teachers to mentor, interpret, and respond, or does it narrow their role to implementation of machine-generated recommendations? These are not technical questions alone. They are questions about the profession itself.

Used responsibly, AI can help educators reclaim time, widen access, and improve responsiveness. Used carelessly, it can erode trust, flatten judgment, and weaken the very forms of learning schools are meant to protect. The goal, therefore, is not to decide whether educators should use AI in the abstract, but to ensure that every use is guided by clear purpose, ethical restraint, and strong human oversight. When those conditions are met, AI can serve education without displacing the humanity at its center.

For quick reference, the following summary rule can guide practice: use AI to support preparation, access, feedback, and reflection; use caution when AI interprets student performance or recommends action; do not use AI to replace human judgment, obscure student authorship, or make high-stakes decisions about learners.

Appendix B:
Human-in-the-Loop Protocol for
Teaching and Assessment

Artificial intelligence can support teaching and assessment in powerful ways, but only when educators remain actively involved in interpretation, decision-making, and ethical oversight. Throughout this book, one principle appears repeatedly: AI should assist educational work, not control it. The purpose of a human-in-the-loop protocol is to make that principle operational. Rather than relying on general cautions or abstract commitments, this appendix provides a practical structure for ensuring that human judgment remains central whenever AI is used in instruction, feedback, assessment, or intervention.

A human-in-the-loop approach means that AI may generate, sort, summarize, recommend, or flag, but it does not make final educational decisions on its own. Teachers, school leaders, and other qualified professionals remain responsible for determining what information matters, how it should be interpreted, whether it is fair, and what action should follow. This is especially important in schools because educational decisions are rarely technical alone. They are relational, contextual, developmental, and ethical. AI may reveal patterns, but it cannot understand a student's full circumstances, intent, effort, emotional state, or history in the way a professional educator can.

This protocol is designed for use in four major areas: instructional planning, classroom feedback, student assessment, and student support or intervention. In each area, the same operational rule applies: AI may inform professional action, but it may not replace professional judgment. Whenever the stakes are high, whenever interpretation is uncertain, or whenever student dignity or fairness may be affected, the human educator must review the output, apply context, and make the final decision.

The protocol begins with a threshold question: Is this a task that can be responsibly supported by AI at all? If the task is low-risk, administrative, or generative in nature, such as drafting lesson options, generating practice questions, translating directions, or summarizing broad performance trends, AI may be used with standard oversight. If the task involves grading, discipline, student risk prediction, academic integrity, individualized intervention, or interpretation of student thinking, then elevated oversight is required. If the task would delegate moral judgment, authorship determination, or a high-stakes student outcome to AI, the use is inappropriate.

The first stage of the protocol is **Define the task clearly**. Before using AI, the educator should identify the exact purpose of the task. Is the goal to save planning time, widen access, identify patterns, generate options, or provide feedback? Vague use invites misuse. If the task has no clear educational purpose, AI should not be introduced. A clearly defined purpose allows the educator to decide whether the tool is being used to support learning or simply to accelerate activity without pedagogical value.

The second stage is **Classify the level of risk**. Low-risk uses include brainstorming lesson ideas, generating examples, reformatting content, or creating optional scaffolds that the teacher will later review. Medium-risk uses include AI-assisted feedback, performance summaries, assessment suggestions, or personalized learning recommendations. High-risk uses include grading recommendations, integrity judgments, intervention flags, behavioral interpretations, and predictive analytics tied to individual students. The higher the risk, the stronger the human review must be. In high-risk cases, no action should be taken until a qualified educator has independently evaluated the recommendation in context.

The third stage is **Review the output for accuracy and appropriateness**. AI-generated content should never be assumed to be correct simply because it is fluent or plausible. Teachers should verify factual claims, check for alignment to standards or learning goals, inspect for bias or omission, and determine whether the tone, complexity, and structure fit the students and context. In assessment-related uses, teachers should ask not only whether the output is correct, but whether it reflects the kind of evidence the task is supposed to capture. If an AI recommendation conflicts with professional judgment or with what the educator knows about the student, the teacher's judgment prevails.

The fourth stage is **Apply contextual interpretation**. This is the heart of the human-in-the-loop model. Educational data and AI recommendations do not interpret themselves. A flagged student may be managing responsibilities outside school. Weak participation may reflect anxiety rather than disengagement. A structurally unusual piece of writing may reflect multilingual development or creative experimentation rather than deficiency. The educator must consider the student's developmental history, cultural and linguistic background, classroom behavior over time, prior performance, and any relevant relational knowledge before accepting or acting on AI-generated output. AI can detect signals; only humans can judge their meaning.

The fifth stage is **Make and document the human decision**. After reviewing the output and interpreting it in context, the educator decides whether to accept, reject, modify, or ignore the AI recommendation. In high-stakes situations, it is useful to document both the AI-generated suggestion and the human rationale for the final decision. This practice protects professional accountability, supports transparency, and creates a record that can be reviewed if questions later arise. It also reinforces the principle that the AI output was advisory, not determinative.

The sixth stage is **Communicate transparently when needed**. Students, families, and colleagues should not be left uncertain about whether AI influenced a decision that affects learning or evaluation. Transparency does not require technical detail in every case, but it does require honesty about the role the system played. If AI was used to generate feedback, recommend groupings, or support preliminary scoring, that role should be explainable in language that is clear and non-defensive. A strong test of ethical use is whether the educator can describe the process openly and still defend it as fair, professional, and educationally sound.

The protocol also requires attention to **nondelegable responsibilities**. Certain educational functions should never be handed over to AI in final form. These include assigning final grades without teacher review, determining disciplinary consequences, deciding whether a student cheated based solely on AI-detection tools, making placement decisions without human interpretation, evaluating emotional state or intent as if the system possessed insight into lived reality, and replacing teacher-written judgment in narrative evaluations or recommendations. These responsibilities are inseparable from professional ethics and cannot be outsourced without compromising fairness and trust.

In teaching, the human-in-the-loop protocol protects instructional integrity. A teacher may use AI to generate draft lesson materials, but must still decide whether those materials are accurate, culturally responsive, developmentally appropriate, and worth teaching. A teacher may use AI to suggest differentiation options but must still determine whether the options match students' needs. A teacher may use AI to summarize class-level error patterns but must still interpret which misconceptions matter and how best to respond. In every case, AI extends teacher reach without displacing teacher expertise.

In assessment, the protocol protects both validity and fairness. A teacher may use AI to identify features of student work, highlight areas for revision, or generate preliminary observations. The teacher must still determine whether the student has demonstrated understanding, whether the task captured authentic evidence of learning, and whether the feedback reflects the goals of the assessment. AI can accelerate response cycles, but it cannot reliably determine originality, nuance, creativity, or growth in the way a thoughtful educator can. The final interpretation of student performance must remain human.

In student support and intervention, the protocol protects dignity. AI systems may flag attendance patterns, behavioral indicators, or academic risk factors, but educators must resist the temptation to treat these signals as conclusions. Predictive tools can be useful prompts for attention, but they should never become labels that narrow expectations or justify predetermined responses. Human educators must examine the context, confer with others when appropriate, and treat the student as more than the profile produced by a system. The protocol therefore insists that any AI-based alert be interpreted as an invitation to investigate, not as a decision already made.

School leaders can use this protocol to establish local expectations for responsible AI use. At a minimum, leaders should identify which tasks are permitted with standard oversight, which require elevated review, and which are prohibited. They should also ensure that staff receive training not only in how to operate AI tools, but in how to question them, verify them, and refuse them when necessary. A school that adopts AI without a human-in-the-loop protocol risks confusing automation with wisdom. A school that adopts AI with such a protocol affirms that technology is valuable only when it remains accountable to educational purpose and human values. The protocol can be summarized in six operational steps:

1. **Define the purpose.** State exactly why AI is being used.

2. **Classify the risk.** Determine whether the use is low-, medium-, or high-risk.

3. **Review the output.** Check for accuracy, bias, alignment, and appropriateness.

4. **Interpret in context.** Apply knowledge of the student, classroom, and school setting.

5. **Make the human decision.** Accept, modify, or reject the recommendation and retain responsibility.

6. **Communicate and document as needed.** Be transparent, especially when outcomes affect students directly.

A final rule should guide all use: the more a decision affects student opportunity, identity, evaluation, or well-being, the less acceptable it is to rely on AI without substantial human review. Efficiency is never sufficient justification for surrendering professional judgment. Human-in-the-loop practice is not a technical preference; it is an ethical requirement for any school that wishes to use AI responsibly while preserving fairness, trust, and humanity at the center of education.

Appendix C:
Assessment Redesign Tool for the Age of AI

Artificial intelligence (AI) has made it necessary for educators to reconsider not only how students complete academic work, but also how schools define credible evidence of learning. When students can use AI tools to generate fluent prose, summarize readings, solve routine problems, and imitate disciplinary conventions with increasing ease, traditional assignments no longer guarantee that the submitted product reflects the student's own understanding. This does not mean assessment is impossible in the age of AI. It means assessment must be redesigned so that it captures reasoning, process, judgment, reflection, and application in ways that are more difficult to outsource and more valuable educationally.

This tool is intended to help teachers, instructional leaders, and teacher preparation programs evaluate existing assessments and redesign them for stronger validity in AI-mediated environments. Its purpose is not to eliminate AI from learning, nor to create a culture of suspicion. Rather, it is to help educators design assessments that preserve authentic evidence of student thinking while still allowing responsible, transparent, and pedagogically justified use of AI where appropriate. Throughout this book, the central principle has remained the same: technology may support learning, but it should not replace the human judgment of the teacher or the intellectual agency of the learner.

An assessment redesign process should begin with a foundational question: What evidence of learning is this task intended to produce? Too often, assignments are built around products rather than evidence. A five-paragraph essay, slide deck, worksheet, or take-home response may appear to measure understanding, but in practice it may only measure whether a student or a machine can assemble a plausible final product. Before redesigning any task, educators should identify the specific knowledge, reasoning, habits of mind, or disciplinary moves they want students to demonstrate. Once the desired evidence is clear, the task can be evaluated for its vulnerability to AI substitution and revised accordingly.

The first dimension of redesign is **evidence clarity**. A strong assessment makes clear what students must show, not simply what they must submit. For example, if the goal is historical reasoning, the evidence should involve sourcing, contextualization, argumentation, and interpretation, not merely the production of a polished essay about a historical topic. If the goal is scientific understanding, the evidence should involve explanation of phenomena, design of investigation, interpretation of data, or justification of conclusions, not merely a completed lab report in conventional format. If the goal is literary analysis, the evidence should involve close reading, inference, and interpretive judgment, not just grammatically polished prose. When teachers clarify the evidence they seek, they are better able to design tasks that reveal actual understanding.

The second dimension is **process visibility**. In the age of AI, final products alone are often insufficient evidence. Stronger assessments make the learning process visible. This can be done through annotated drafts, planning notes, revision logs, think-alouds, checkpoints, oral explanations, conference discussions, version histories, process reflections, or in-class stages of work. A visible process allows teachers to see how ideas were formed, revised, and defended. It also makes it easier to distinguish between support that strengthens learning and substitution that obscures it. Process visibility does not require surveillance. It requires thoughtful design that values the path of thinking, not only the polished endpoint.

The third dimension is **cognitive demand**. Assessments that rely heavily on generic summary, routine explanation, predictable paragraph structures, or formulaic problem types are more easily completed by AI without deep understanding. By contrast, tasks that require comparison, interpretation, application to unfamiliar contexts, oral defense, local relevance, synthesis across sources, or explanation of reasoning are more likely to elicit authentic thinking. This does not mean every assessment must be elaborate. It means the teacher should ask whether the task requires students to think in ways that matter, or merely to produce something that looks academic. Assessments should privilege judgment, transfer, and disciplined reasoning over surface completion.

The fourth dimension is **context specificity**. AI performs best on generalized academic tasks. Assessments become more robust when they are anchored in local, personal, classroom-based, or context-specific conditions that require students to draw on shared experiences, class texts, discussion history, field observations, lab results, school-based data, or disciplinary decisions made during instruction. A student asked to write a generic essay on climate change can easily turn to AI. A student asked to analyze a local environmental dataset, connect it to class discussion, explain their reasoning in conference, and revise based on teacher questioning is demonstrating a more defensible form of learning. Context specificity increases authenticity by tying the task to actual learning experiences rather than generic prompts.

The fifth dimension is **AI transparency**. Not every use of AI is inappropriate. In some cases, students may appropriately use AI for brainstorming, language support, feedback, translation, organizational help, or revision assistance. The issue is not mere use, but undisclosed or inappropriate substitution. Redesigned assessments should therefore include explicit directions about whether AI use is permitted, prohibited, or limited to certain stages. They should also require disclosure when AI has been used in a meaningful way. A short reflection or disclosure statement can help teachers understand what support was used, why it was used, and how the student remained intellectually responsible for the work. Transparency is more educationally productive than guesswork or punitive detection culture.

The sixth dimension is **teacher interpretive authority**. Assessments must be designed so that the teacher remains able to evaluate quality, growth, and understanding using professional judgment. AI-generated scoring suggestions, analytic dashboards, or automated writing feedback may help inform the process, but they should not replace the teacher's interpretation of what the student has demonstrated. A redesigned assessment should preserve opportunities for the educator to question, probe, compare, and contextualize student performance. This is especially important in cases involving creativity, nuance, originality, multilingual expression, or unconventional but meaningful approaches to the task. The tool that follows can be used to review any assignment.

Assessment Redesign Review Questions:

1. **What learning is this task supposed to measure?** State the specific knowledge, reasoning, skill, or disposition the assessment is meant to capture.

2. **What evidence would convincingly show that learning?** Identify the actual behaviors, explanations, decisions, or products that would demonstrate understanding.

3. **Could AI complete the task successfully without the student demonstrating that evidence?** If the answer is yes, the task needs revision.

4. **Does the assessment make student thinking visible?** Look for places to add process notes, oral explanation, drafts, reflection, checkpoints, or live application.

5. **Does the task require reasoning, interpretation, application, or judgment?** If it primarily rewards polished output or routine summary, raise the cognitive demand.

6. **Is the task tied to local classroom context or shared learning experiences?** If not, consider using class-specific content, discussion references, local data, or teacher-guided materials.

7. **Are expectations for AI use explicit?** State clearly whether AI is allowed, how it may be used, and what disclosure is required.

8. **Does the teacher remain the final evaluator of student understanding?** If the assessment relies too heavily on automated scoring or AI interpretation, redesign is needed.

Teachers can also use a simple three-level redesign screen:

- A task is **highly vulnerable** if AI can produce a satisfactory final submission with minimal student thinking, if process is invisible, and if the teacher has little basis for verifying understanding beyond the submitted product.

- A task is **moderately vulnerable** if AI could assist substantially, but the teacher still has some visibility into process, context, or explanation.

- A task is **more defensible** if the task requires student reasoning in context, includes visible process, clarifies AI boundaries, and preserves teacher judgment in evaluation.

Several redesign moves can strengthen existing assessments without requiring complete replacement.

- One useful move is to **shift from product-only assessment to product-plus-process assessment.** Instead of grading only the final essay, teachers might collect the proposal, outline, draft, revision notes, and short reflection. Instead of grading only the completed lab report, they might include a rationale for method choices, a conference discussion, or a claim-evidence-reasoning explanation completed in class. This approach makes intellectual work more visible and reduces the chance that a polished final product will mask weak understanding.

- Another move is to **add oral or live components**. A brief conference, oral defense, presentation segment, or question-and-answer exchange can provide powerful evidence of understanding. Students need not perform extensively for every assessment, but selective oral elements allow teachers to probe whether students can explain, justify, and elaborate on the work they submitted. This is especially valuable when the written product alone does not provide enough confidence about authorship or depth of understanding.

- A third move is to **increase the role of application and transfer**. AI is strong at reproducing conventional forms, but deeper learning is better revealed when students apply ideas to unfamiliar scenarios, local cases, classroom-generated data, real audiences, or evolving problems. Assessments become more robust when students must do something with knowledge rather than merely restate it.

- A fourth move is to **build in reflection on tool use**. If students are allowed to use AI in limited ways, ask them to explain what they used it for, what they accepted or rejected, and how they ensured that the final work still reflected their own thinking. This turns AI use into an object of metacognitive reflection rather than a hidden shortcut. It also helps students learn responsible habits of disclosure, critique, and self-regulation.

- A fifth move is to **redesign rubrics to value thinking, not just polish**. In the age of AI, surface features such as fluency, grammatical correctness, or generic organizational neatness may no longer be dependable indicators of understanding. Rubrics should place stronger emphasis on reasoning, interpretation, specificity, decision-making, conceptual accuracy, use of evidence, and responsiveness to context. This helps ensure that what is rewarded is the student's intellectual work rather than merely the smoothness of the finished product.

The redesign tool can be applied across disciplines.

- In English language arts, a traditional literary essay can be redesigned by requiring students to annotate passages in class, explain interpretive choices in conference, and submit a reflection on how their interpretation changed across drafts.

- In social studies, a generic research report can be redesigned into a document-based argument tied to class-selected sources, followed by an oral explanation of sourcing and contextualization decisions.

- In science, a standard lab report can be strengthened by adding a design rationale, error analysis, and live explanation of how the student interpreted data.

- In mathematics, students can be asked not only to solve, but to explain why a strategy works, compare methods, and apply concepts to a new or locally relevant problem.

- In arts and design fields, students can document iterations, explain aesthetic decisions, and show how feedback shaped revision.

This tool is also useful for school leaders and program designers. Departments can use it to audit common assignments, identify areas where traditional tasks no longer produce credible evidence, and support teachers in redesigning assessments collaboratively. Teacher preparation programs can use it to help future educators understand that assessment design is now inseparable from questions of AI literacy, instructional validity, and academic integrity. In both settings, the redesign conversation should move away from policing and toward better design.

The goal of assessment redesign is not to outmaneuver technology for its own sake. It is to preserve meaningful evidence of student learning in a world where polished products can increasingly be generated by tools. Strong assessments remain possible, but they must now be built with greater intentionality. They must reveal reasoning, make process visible, define legitimate support clearly, and keep the teacher in the role of professional interpreter. When educators redesign assessment in these ways, they do more than respond to AI. They strengthen the quality, fairness, and authenticity of learning itself.

For quick use, the tool can be summarized in one guiding rule: redesign any assessment that can be completed successfully by AI without requiring visible student reasoning, contextual application, or teacher-verifiable evidence of learning.

Appendix D
School or District AI Readiness Rubric

This appendix is designed to help school and district leaders assess whether they are prepared to adopt artificial intelligence in ways that are educationally sound, ethically responsible, and operationally sustainable. It is not intended as a compliance checklist alone. Rather, it is a reflective planning tool that helps leaders determine whether the conditions for responsible implementation are in place before AI use expands across classrooms, assessment systems, communication platforms, or decision-making processes. The rubric reflects a central argument of this book: successful AI integration depends less on the power of the tool than on the quality of the human systems guiding its use.

Schools and districts can use this rubric in several ways. Leadership teams may complete it collaboratively as part of strategic planning. Principals may use it with faculty committees to identify building-level strengths and gaps. District leaders may use it to compare readiness across schools and to sequence implementation more deliberately. Teacher preparation programs and professional learning teams may also find it useful as a discussion protocol for examining how policy, pedagogy, ethics, and equity intersect in AI adoption.

The rubric is organized around eight domains that are essential to responsible AI integration: vision and purpose, governance and policy, privacy and data protection, equity and access, instructional readiness, assessment integrity, professional learning, and monitoring and continuous improvement. Each domain includes four performance levels: Emerging, Developing, Operational, and Transformative. These levels are not intended to function as labels of success or failure. They are indicators of institutional readiness and capacity. Most schools will find that they are operating at different levels across domains. That is expected. The value of the rubric lies in helping leaders identify where further planning, communication, or support is needed before implementation deepens.

Schools or districts should begin by assembling a representative review team. This team should ideally include administrators, teachers, instructional coaches, technology staff, student support personnel, and, when appropriate, student or family representatives. Team members should rate each domain using available evidence, such as board policies, acceptable-use guidelines, professional development records, vendor contracts, assessment practices, classroom observations, and stakeholder feedback. After each domain is rated, the team should identify priorities for action, assign responsibility, and determine a timeline for follow-up review.

A school or district is rarely ready for broad AI implementation if it lacks clarity of purpose, clear policies, teacher support, or safeguards for students. For that reason, the rubric should not be used to justify rapid adoption. It should be used to strengthen thoughtful, transparent, and equitable implementation.

Using the Rubric

After reviewing the domains and their associated descriptors, school or district leadership teams should select the level that most closely reflects current practice. Teams should base ratings on available evidence rather than perceptions or aspirations. The goal is not to achieve the highest possible score in every category, but to develop an accurate understanding of current readiness and identify areas for improvement.

After completing all domains, review results collectively to identify strengths, gaps, and priorities for action. Attention should be given to domains where implementation activity exceeds institutional capacity or where important safeguards have not yet been established.

Domain 1. Vision and Purpose

- **Emerging:** AI use is reactive, inconsistent, or driven mainly by curiosity, vendor marketing, or individual experimentation. No clear rationale has been established for how AI supports institutional goals.

- **Developing:** Leaders have begun discussing possible AI uses and have identified some early goals, but the rationale remains broad, fragmented, or unevenly understood.

- **Operational:** The school or district has a clear purpose for AI use connected to instructional improvement, student support, operational efficiency, or equitable access. Stakeholders can explain its intended role.

- **Transformative:** AI use is guided by a well-communicated, mission-aligned vision centered on human judgment, student dignity, and educational purpose. Adoption and evaluation decisions consistently reflect this vision.

Domain 2. Governance & Policy

- **Emerging:** No formal AI policy exists. Expectations for staff and students are unclear or absent. Decision-making is informal and uncoordinated.

- **Developing:** Initial guidance has been drafted, but policies are limited in scope or inconsistently communicated. Questions about acceptable use, accountability, and oversight remain unresolved.

- **Operational:** Formal policies or procedures guide AI use by staff and students. These address acceptable use, approval processes, oversight responsibilities, and limits on automation in high-stakes contexts.

- **Transformative:** Governance structures are transparent, regularly reviewed, and responsive to changing technologies. Policy reflects a clear commitment to human-in-the-loop decision-making, ethical use, and accountability.

Domain 3: Privacy and Data Protection

- **Emerging:** AI tools may be used without systematic review of privacy implications. Staff may not understand what data are collected, stored, or shared.

- **Developing:** Privacy concerns are recognized, and some tools are reviewed for compliance, but vetting and communication procedures remain incomplete or inconsistently applied.

- **Operational:** AI tools are subject to a clear vetting process that includes review of data collection, storage, access, retention, and vendor terms. Policies align with legal and institutional requirements.

- **Transformative:** Data governance is proactive, transparent, and rigorous. Privacy protections are embedded in procurement, implementation, and communication. Safeguards are regularly audited and strengthened.

Domain 4: Equity and Access

- **Emerging:** AI use is uneven and may be limited to better-resourced classrooms or programs. Little attention has been given to bias, accessibility, or disparate impact.

- **Developing:** Leaders recognize equity concerns and have begun considering access, representation, and inclusion. Some supports exist, but implementation remains uneven.

- **Operational:** Deliberate steps have been taken to ensure broad access to AI-related tools and learning opportunities. Equity considerations inform procurement, deployment, accessibility, and support for diverse learners.

- **Transformative:** Equity is treated as a primary lens for all AI-related decisions. The institution actively monitors for bias, addresses disparities in access and outcomes, and uses AI to expand opportunity for historically marginalized students.

Domain 5: Instructional Readiness

- **Emerging:** Teachers are experimenting with AI individually, but there is little shared guidance on instructional quality, appropriate uses, or pedagogical alignment.

- **Developing:** Teachers have begun using AI for planning, differentiation, or support, and some examples of effective practice exist, but expectations remain unclear or uneven.

- **Operational:** Teachers have clear guidance for using AI in ways that support standards, learning goals, differentiation, and student agency. AI supports sound pedagogy rather than substituting for it.

- **Transformative:** Instructional AI use reflects a mature understanding of pedagogy, ethics, and human judgment. Teachers use AI to strengthen inquiry, accessibility, personalization, and reflection while preserving authentic student thinking.

Domain 6: Assessment Integrity

- **Emerging:** The impact of AI on assessment has not been addressed systematically. Existing assignments and integrity expectations may no longer align with current technologies.

- **Developing:** Staff have begun discussing acceptable and unacceptable uses of AI in student work. Some assessment redesign is underway, but practices remain inconsistent.

- **Operational:** Expectations for AI disclosure, academic integrity, and assessment design have been established. Teachers are revising tasks to emphasize process, reasoning, reflection, and authentic evidence of learning.

- **Transformative:** Assessment policy and practice have been thoughtfully redesigned for an AI-mediated environment. The institution has built a culture of transparency, credible evidence, and human judgment rather than relying on detection alone.

Domain 7: Professional Learning and Capacity Building

- **Emerging:** Few staff have received formal support related to AI. Knowledge is uneven, and confidence is low or highly variable.

- **Developing:** Introductory professional learning has begun, often through workshops or optional sessions. Awareness is increasing, but deeper pedagogical and ethical competence remains limited.

- **Operational:** Professional learning is ongoing, job-embedded, and connected to staff roles. Educators and leaders are building practical competence in AI literacy, ethical use, instructional integration, and policy awareness.

- **Transformative:** Professional learning is sustained, collaborative, and strategically aligned to institutional goals. Staff function as competent users, critical evaluators, and ethical decision-makers who contribute to institutional learning.

Domain 8: Monitoring, Evaluation, and Continuous Improvement

- **Emerging:** AI implementation is not being systematically monitored. Success is assumed rather than evaluated.

- **Developing:** Some feedback is being collected, and leaders are beginning to identify implementation issues, but evaluation remains informal or incomplete.

- **Operational:** The school or district has a process to review AI use, gather stakeholder feedback, assess outcomes, and make ongoing improvements.

- **Transformative:** The school or district analyzes data annually to ensure that policies and programs remain current and effective.

Scoring and Interpretation

After completing the rubric, teams should identify their approximate level in each domain and then step back to look for patterns. A school or district that rates itself highly in instructional experimentation but weakly in governance and privacy is not fully ready for scaled implementation. Likewise, a district with strong policies but limited professional learning may have compliance structures in place without the staff capacity to enact them well. Readiness is not defined by isolated strengths. It is defined by coherence across systems.

Schools or districts that are mostly at the Emerging level should focus first on foundational planning. This includes clarifying purpose, pausing uncoordinated expansion, and establishing basic policy and privacy safeguards. Those at the Developing level should concentrate on building consistency across schools, clarifying expectations, and deepening staff capacity. Those at the Operational level are generally positioned for thoughtful implementation but should continue monitoring equity, ethics, and impact. Those approaching the Transformative level should not assume the work is complete. Rather, they should view readiness as an ongoing institutional responsibility requiring continued reflection, review, and adaptation.

Planning Next Steps

Once the rubric is completed, leadership teams should identify two or three priority actions for the next phase of implementation. Those actions might include drafting or revising AI-use policies, strengthening vendor-review protocols, redesigning assessments, expanding professional development, or improving communication with families. Teams should also identify what evidence will be used to measure progress and when the rubric will be revisited.

Suggested reflection questions include: What domain presents the greatest risk if left unaddressed? Where are we moving faster than our systems are prepared to support? Which stakeholders have not yet had meaningful input into our AI planning? What evidence do we have that AI use is advancing equity and learning rather than simply increasing efficiency? How will we know whether our implementation is strengthening, rather than weakening, the human dimensions of education?

Used well, this rubric helps leaders move beyond enthusiasm or fear toward disciplined, ethical decision-making. Artificial intelligence should not be adopted simply because it is available, nor rejected simply because it is unfamiliar. It should be evaluated according to whether the institution is prepared to guide it wisely. In that sense, readiness is not merely technical. It is educational, ethical, and organizational. The ultimate question is not whether a school or district is using AI, but whether it is ready to do so in a way that protects human dignity, strengthens teaching, and supports meaningful learning.

Appendix E:
Sample School AI Use Policy

This sample policy is designed for adaptation by schools or districts seeking to establish clear expectations for the responsible use of artificial intelligence in teaching, learning, assessment, and operations. It is intentionally written in formal policy language so it can serve as a starting point for local revision. Schools should review it considering applicable state law, district procedures, collective bargaining agreements, privacy requirements, and existing academic integrity policies.

Purpose:

The purpose of this policy is to establish clear expectations for the ethical, educationally sound, and responsible use of artificial intelligence in the school environment. Artificial intelligence can support teaching, learning, accessibility, communication, and operational efficiency. At the same time, its use raises important concerns related to privacy, accuracy, bias, transparency, academic integrity, and human judgment. This policy is intended to ensure that AI is used in ways that strengthen learning, protect student dignity, and preserve the professional responsibility of educators.

Guiding Principles:

The school/district affirms that artificial intelligence must serve educational goals rather than define them. Human judgment remains central to all significant instructional, assessment, disciplinary, and student-support decisions. AI tools may assist educators and students, but they do not replace the professional expertise of teachers, the supervisory role of school leaders, or the intellectual responsibility of learners. All AI use within the school/district shall be guided by the following principles:

- Human oversight. School/district personnel remain responsible for reviewing, interpreting, and approving significant decisions influenced by AI.

- Educational purpose. AI tools may be used only when they support legitimate educational, accessibility, administrative, or communication needs.

- Equity and inclusion. AI use must not widen disparities in access, opportunity, or treatment, and schools must remain alert to the possibility of algorithmic bias.

- Privacy and security. Student and staff data must be protected in accordance with all applicable laws, policies, and contractual safeguards.

- Transparency. Students, staff, and families should be informed when AI is used in meaningful ways that affect instruction, feedback, communication, or decision-making.

- Academic integrity. AI may support learning, but it may not be used to misrepresent authorship, bypass required thinking or undermine authentic evidence of student learning.

Scope

This policy applies to all students, employees, contractors, and approved third-party providers who use artificial intelligence tools in connection with school-related work, instruction, communication, assessment, or operations. It applies whether AI tools are accessed on school-owned devices, personal devices used for school purposes, district-approved platforms, or authorized third-party applications.

Definition

For purposes of this policy, artificial intelligence refers to digital systems that can generate, analyze, classify, predict, recommend, summarize, translate, score, or otherwise process information in ways that simulate aspects of human reasoning, language, pattern recognition, or decision support. This includes, but is not limited to, generative AI systems, automated writing tools, adaptive learning platforms, predictive analytics systems, image or media generators, automated scoring systems, and AI-supported chat or tutoring applications.

Approved and Prohibited Uses

School/district personnel and students may use approved AI tools for legitimate school purposes, including instructional planning, accessibility support, translation, differentiated materials, formative feedback, administrative drafting, and other educational uses consistent with this policy and with school or district guidance. Artificial intelligence may not be used for any of the following purposes:

- To make final high-stakes decisions about grading, discipline, promotion, retention, placement, or student support without meaningful human review.

- To submit work that falsely represents AI-generated content as entirely the student's own original work.

- To enter confidential, personally identifiable, or protected student information into unapproved AI tools.

- To generate or distribute harmful, discriminatory, harassing, sexually inappropriate, deceptive, or unsafe content.

- To surveil, profile, or classify students in ways that are inconsistent with law, policy, or educational ethics.

- To replace teacher evaluation of student learning in cases where professional judgment is required.

Staff Responsibilities

All school/district personnel are responsible for using AI tools carefully, professionally, and in a manner consistent with school values. Staff members who use AI in their work shall:

- Use only school- or district-approved tools when student, staff, or school data are involved.

- Review AI-generated outputs for accuracy, appropriateness, bias, and alignment before use.

- Protect student privacy and avoid entering confidential information into unapproved systems.

- Inform students when AI is being used in instruction, feedback, or classroom activities in ways that materially affect the learning process.

- Use AI to support, not replace, sound pedagogy, professional judgment, and human relationships.

- Design assessments and learning experiences that preserve authentic student thinking and credible evidence of learning.

- Report concerns related to AI errors, bias, privacy, misuse, or policy violations to school administration.

Student Responsibilities

Students may use AI tools only as permitted by their teacher, school, or district. When AI use is authorized, students are expected to use these tools responsibly, transparently, and in ways that support learning rather than replace it. Students shall:

- Follow teacher directions regarding when and how AI use is permitted.

- Disclose AI use when required by the teacher, assignment, or school policy.

- Use AI as a support for brainstorming, clarification, feedback, revision, or approved assistance, not as a substitute for their own reasoning when original work is expected.

- Verify information generated by AI rather than assuming it is accurate.

- Avoid submitting AI-generated work as their own when such use is prohibited.

- Refrain from using AI to cheat, plagiarize, impersonate others, generate harmful content, or violate school rules.

Academic Integrity

The school/district recognizes that AI changes how students access support, draft ideas, and revise work. For that reason, academic integrity expectations must remain clear. Teachers shall communicate assignment-specific expectations regarding acceptable and unacceptable AI use. In general, students may use AI only in ways explicitly permitted by the teacher. When AI contributes to student work, students may be required to disclose the tool used, the purpose for which it was used, and the extent to which its output influenced the final product.

Unauthorized use of AI to complete assignments, generate responses, or conceal the absence of original student thinking may be treated as academic dishonesty under the school's/district's existing integrity policy. Consequences shall be determined in accordance with established school/district procedures and with attention to student age, developmental level, and the nature of the violation.

Assessment and Human Judgment

Artificial intelligence may assist with aspects of assessment, such as generating practice questions, offering preliminary feedback, identifying patterns in student responses, or supporting formative assessment. However, teachers remain responsible for evaluating student work, interpreting evidence of learning, and making final judgments about academic performance. Automated systems may inform but shall not replace teacher judgment in consequential assessment decisions.

Teachers are encouraged to design assessments that emphasize process, reasoning, reflection, revision history, discussion, performance, and other forms of authentic evidence when appropriate in an AI-mediated environment.

Privacy and Data Protection

The school/district shall take reasonable steps to ensure that AI tools used for school/district purposes comply with applicable privacy laws and district data governance requirements. Student information may be entered only into approved systems that have undergone appropriate review for security, data retention, vendor access, and legal compliance.

The school/district shall not require students to use AI tools that collect personal data without appropriate review and authorization. When required by law or policy, the school shall provide notice and obtain consent for specific uses of AI-related systems.

Equity, Accessibility, and Bias

The school/district recognizes that AI tools can both support and undermine equity. AI shall be used in ways that expand access for multilingual learners, students with disabilities, and other students who benefit from adaptive or assistive supports. At the same time, the school shall remain alert to the possibility that AI systems may reflect cultural, linguistic, racial, gender, or socioeconomic bias.

School/district personnel shall monitor AI-supported practices for unfair impact, inappropriate recommendations, or exclusionary effects. Concerns related to bias or inequity shall be reviewed promptly and used to inform future decisions about tool selection, implementation, or discontinuation.

Communication with Families

The school/district shall communicate clearly with families about the school's approach to AI use. This communication may include the educational purposes for which AI is used, the safeguards in place to protect privacy, the expectations for student use, and the processes for raising questions or concerns. Families should be able to understand both the opportunities and the limitations associated with AI in the school setting.

Professional Learning

The school/district shall provide ongoing professional learning to support responsible AI use by faculty and staff. This learning should address AI literacy, privacy, ethics, academic integrity, instructional design, accessibility, and bias awareness. Staff should not be expected to implement AI well without adequate guidance, support, and time for reflection.

Monitoring and Review

School/district administration shall monitor the implementation of this policy and review it periodically considering emerging technologies, legal developments, stakeholder feedback, and observed practice. Revisions should be made as needed to ensure continued alignment with educational priorities, student well-being, and ethical use.

Policy Violations

Violations of this policy by students shall be addressed in accordance with the school's/district's student code of conduct, academic integrity procedures, and disciplinary guidelines. Violations by employees shall be addressed in accordance with district policy, professional expectations, and applicable employment procedures. Misuse by vendors or contractors may result in suspension or termination of access, contract review, or other appropriate action.

Conclusion

Artificial intelligence can be a useful educational support when guided by clear expectations, strong professional judgment, and a commitment to human dignity. This policy affirms that the role of AI in schools is to assist, not replace, the human relationships, ethical reasoning, and intellectual work at the center of teaching and learning.